2nd International Conference on Reliability Engineering (ICRE 2017)

IOP Conference Series: Materials Science and Engineering
Volume 351

Milan, Italy
20 – 22 December 2017

ISBN: 978-1-5108-6282-1
ISSN: 1757-8981

Printed from e-media with permission by:

Curran Associates, Inc.
57 Morehouse Lane
Red Hook, NY 12571

proceedings.com

Some format issues inherent in the e-media version may also appear in this print version.

Printed by Curran Associates, Inc. (2018)

For permission requests, please contact the Institute of Physics
at the address below.

Institute of Physics
Dirac House, Temple Back
Bristol BS1 6BE UK

Phone: 44 1 17 929 7481
Fax: 44 1 17 920 0979

techtracking@iop.org

Additional copies of this publication are available from:

Curran Associates, Inc.
57 Morehouse Lane
Red Hook, NY 12571 USA
Phone: 845-758-0400
Fax: 845-758-2633
Email: curran@proceedings.com
Web: www.proceedings.com

Table of contents

Volume 351

2017 2nd International Conference on Reliability Engineering (ICRE 2017)

20–22 December 2017, Milan, Italy

Accepted papers received: 16 April 2018 - Published online: 14 May 2018

Engineering Maintenance and Quality Management

Information Technology and Security

PREFACE

Welcome to 2017 2nd International Conference on Reliability Engineering (ICRE 2017) held in Milan, Italy during Dec 20-22, 2017. The aim objective of ICRE 2017 is to provide a platform for researchers, engineers, academicians as well as industrial professionals from all over the world to present their research results and development activities in Reliability Engineering. This conference provides opportunities for the delegates to exchange new ideas and application experiences face to face, to establish business or research relations and to find global partners for future collaboration.

The present volume contains the selected papers submitted to this conference and provides up-to-date, comprehensive and worldwide state-of-the art knowledge in the field of Reliability Engineering. All papers included in this collection had undergone the careful peer-review by the conference technical committee members and international reviewers before they're selected for publication. Studies presented in this volume cover the following topics: Equipment Monitoring and Reliability Analysis, Product Design and Reliability Testing, Circuit System Analysis and Detection, Engineering Maintenance and Quality Management, Information Technology and Security.

We'd like to express our heartful gratitude to all the committee members for their great professionalism and efforts. And thank all the sponsors and participants for their valuable contributions and support to ICRE 2017.

ICRE2017
Conference Committee

IOP Conf. Series: Materials Science and Engineering **351** (2018) 011002 doi:10.1088/1757-899X/351/1/011002

Conference Committee

International Advisory Committee

Prof. Terje AVEN, University of Stavanger (UiS), Norway

Conference General Chair

Prof. Enrico Zio, Politecnico di Milano, Italy

Conference Co-chair

Prof. Michael Pecht, University of Maryland, USA

Prof. Rui Kang, Beihang University, China

Steering Committee

Prof. Min Xie, City University of Hong Kong, Hong Kong

Prof. TANG Loon Ching, National University of Singapore, Singapore

Program Chair

Prof. Dr. Francesca Saglietti, University of Erlangen-Nuremberg, Germany

Prof. Jin Wang, Liverpool John Moores University, UK

Prof. Francesco Di Maio, Politecnico di Milano, Italy

Prof. Yi Ding, Zhejiang University, China

Technical Committee

Prof. Zitouni Beidouri, ESTC, Morocco

Prof. Tadashi Dohi, Hiroshima Univ, Japan

Prof. Wei-Chang Yeh, National Tsing Hua University, Taiwan

IOP Conf. Series: Materials Science and Engineering **351** (2018) 011003 doi:10.1088/1757-899X/351/1/011003

Peer review statement

All papers published in this volume of *IOP Conference Series: Materials Science and Engineering* have been peer reviewed through processes administered by the proceedings Editors. Reviews were conducted by expert referees to the professional and scientific standards expected of a proceedings journal published by IOP Publishing.

IOP Conf. Series: Materials Science and Engineering **351** (2018) 012001 doi:10.1088/1757-899X/351/1/012001

Towards generating ECSS-compliant fault tree analysis results via ConcertoFLA

B Gallina[1] Z Haider[1] and A Carlsson[2]

[1] IDT, Mälardalen University, Västerås, Sweden
[2] OHB Sweden, Stockholm (Kista), Sweden

barbara.gallina@mdh.se

Abstract. Attitude Control Systems (ACSs) maintain the orientation of the satellite in three-dimensional space. ACSs need to be engineered in compliance with ECSS standards and need to ensure a certain degree of dependability. Thus, dependability analysis is conducted at various levels and by using ECSS-compliant techniques. Fault Tree Analysis (FTA) is one of these techniques. FTA is being automated within various Model Driven Engineering (MDE)-based methodologies. The tool-supported CHESS-methodology is one of them. This methodology incorporates ConcertoFLA, a dependability analysis technique enabling failure behavior analysis and thus FTA-results generation. ConcertoFLA, however, similarly to other techniques, still belongs to the academic research niche. To promote this technique within the space industry, we apply it on an ACS and discuss about its multi-faceted potentialities in the context of ECSS-compliant engineering.

1. Introduction

Attitude Control Systems (ACSs) play an important role within satellites. More specifically, ACSs contribute to maintaining a certain attitude (i.e., orientation of the satellite in three-dimensional space). Controlling the attitude is necessary to enable satellites (space crafts, which, typically, orbit earth [1]) to accomplish their mission including the fulfillment of their pointing requirements, often referred to as pointing modes [2].

ACSs need to be engineered in compliance with ECSS standards and need to ensure a certain degree of dependability. Thus, dependability analysis is conducted at various levels and by using different techniques. Fault Tree Analysis (FTA) is one of these techniques. This analysis technique is being automated within various Model Driven Engineering (MDE)-based methodologies. The tool supported CHESS-methodology [3] is one of them. This tool-supported methodology incorporates ConcertoFLA [4], a dependability analysis technique enabling failure behavior and propagation analysis and thus FTA-results generation. ConcertoFLA, however, similarly to other techniques, belongs to the academic research niche. To promote this technique within the space industry, extensive explorations are needed. In this paper, we contribute to perform such explorations. More specifically, we exploit the tool-supported CHESS methodology to model and analyze (via ConcertoFLA) an ACS. We also discuss how analysis results can be used to generate ECSS-FTA compliant analysis results.

The rest of the paper is organized as follows. In Section 2, we provide background information. In Section 3, we provide an overview of the entire approach. In Section 4, we focus on the modelling of our ACS. In Section 5, we perform the analysis. In Section 6, we elaborate on FTA generation, based on ConcertoFLA analysis results. In Section 7, we discuss related work. Finally, in Section 8, we present our concluding remarks and future work.

2. Background

In this section, we present the background information on which we base our work.

2.1. *ACSs in Sun Pointing mode*

As mentioned in the introduction, ACSs control the satellite's attitude. This control-related functionality is relative to a frame of reference depending upon the pointing requirement, which is either mission mode or safehold mode [2]. The former refers to the main objective of the satellite and latter contributes to fail safe. For instance, a sun pointing mode could be a safehold mode for a satellite pointing its solar arrays towards the sun to power the critical parts of the satellite. In sun pointing mode, the attitude of a satellite, relative to the sun, is maintained by controlling the torques applied to the satellite by actuator thrusters. A sun sensor measures the angle of the sun beam, in sensor's reference frame, when the beam hits the sensor. The direction of the sun is acquired from this angle and fed into the control system as an input. The control system calculates the torque based on minimizing the pointing error function.

Typically, an ACS (focus on software) is a composite component composed of the following four software components: (1) PDController takes the Sun direction and angular velocity of the satellite in three axes as an input, computes the pointing error and calculates proportional and derivative torques. (2) SteerController also computes the proportional torque by minimizing the pointing error. But, the objective is to compute relatively greater torques for faster convergence to target pointing position. This is important in cases when the satellite is significantly diverging from the desired position e.g., when it is upside down. (3) FeedforwController computes additional torques to achieve the equilibrium state and stabilize the satellite body. A satellite in the space is subject to the disturbance torques coming from outside of the boundary of the satellite. For example, the Sun radiation pressure, gravitational pull of the celestial bodies and the earth atmospheric pressure etc. FeedforwController takes these torques as an input and calculates a complementary torque to counteract the disturbances. Typically, these disturbances torque are calculated by the angular rates of the satellite measured a gyroscope sensor. (4) TorqueSelector takes the torques computed by the above-mentioned controllers, and, based on the current attitude of the satellite, allows a specific torque to be applied on the satellite through the thrusters. If the satellite is oriented 180 degrees opposite to the desired pointing direction, a rapid change in attitude is required. Hence, relatively greater torques, which are computed by SteerController, are applied. On the other hand, when the satellite is closer to the desired pointing direction, rather a seamless convergence and stabilization is of interest. Therefore, smaller torques and stabilization torques, computed by PDController and FeedforwController respectively, are applied to satellite.

2.2. *ECSS-compliant FTA*

ECSS-Q-ST-30 [5], the ECSS standard targeting product assurance, lists the methods for performing dependability analysis, required at all levels of the space system. Fault Tree Analysis (FTA) is a top down approach to assess the reliability of a system, by analyzing the failure state and constructing a probable cause tree. The objective of the analysis is to "ensure that the design conforms to the failure tolerance requirements for combinations of failures" [5]. According to ECSS-Q-ST-30, a system/sub-system level analysis is required to identify the combination of the events that lead to a failure e.g. loss of mission [5]. The end consequence of a failure is defined as the severity, and could be catastrophic, critical, major and minor. The severity level of a failure together with the probability of its occurrence defines the criticality of the event. The severity level of the system in consideration i.e., ACS in Sun pointing mode is classified as critical, since the consequence is loss of the mission.

ECSS-Q-ST-40-12C [6] provides the guidelines for performing ECSS-compliant FTA, required for all the systems/sub-systems with severity level catastrophic, critical or major consequences. For procedural guidelines of performing FTA, ECSS-Q-ST-40-12C refers to IEC-61025 [7]. Additionally, it is suggested that Fault Tree Handbook [8] can be used to complement the standard. IEC-61025 uses general terms to describe FTA analysis, steps to perform the analysis, list of assumptions and standardized shapes and symbols. IEC-61025 requirements are generic and are not limited to space domain. A compliance artifact, as per IEC-61025, is required to include the analysis, data, used symbols, and common cause failures & minimal cut sets; where appropriate. Section 4 discusses these requirements in ConcertoFLA context and their conformance.

2.3. *SafeConcert and ConcertoFLA*

SafeConcert [9] is a meta-model implemented within the CHESS toolset [3] enabling dependability architects to model dependability's information necessary to conduct dependability analysis. SafeConcert is a subset of CHESS-ML (which in turn is an extension of SySML [10]), the meta-model used in CHESS toolset to enable component-based systems design.

ConcertoFLA [4], which builds on top of CHESSFLA [11], allows users (system architects and dependability engineers) to decorate component-based architectural models (specified using CHESS-ML) with dependability-related information, execute Failure Logic Analysis (FLA) techniques, and get the results back-propagated onto the original model. Different FLA techniques are available in the literature [12], and can be used at the early stages of the design phase to achieve a robust architecture with respect to linear relationships. ConcertoFLA builds on top of Failure Propagation Transform Logic (FPTC) [13]. Similar to FPTC, ConcertoFLA is a compositional technique to qualitatively assess the dependability of component-based systems, and partially combines and automatize traditional safety analysis techniques (i.e., FMEA and FTA). ConcertoFLA allows users to calculate the behavior at system-level, based on the specification of the behavior of individual components.

In ConcertoFLA terms, a component can act in four different possible ways (1) source of the failure thus generating a failure due to internal fault, (2) sink of the failure thus avoiding the propagation of the external fault (failure in input) through fault tolerance, (3) propagator of the failure, and (4) transformer of the failure into a different type. ConcertoFLA rules are logical expressions, which specify the component's behavior by describing the input/output relationship. In [14], the syntax for writing the logical expressions is given (subset of the FlaMM meta model). This syntax represents a further refinement of what was presented in [4].

3. Approach overview

Figure 1 shows a high-level view of the overall approach. This approach builds on top of the tool-supported technique presented in [4] and then further developed in [14]. In this paper, we use such approach and customize it in the context of ECSS standards to enable architects and dependability managers to calculate the failure behavior at sub/system level, as well as generating FTA based on the causality paths traced by the ConcertoFLA plugin.

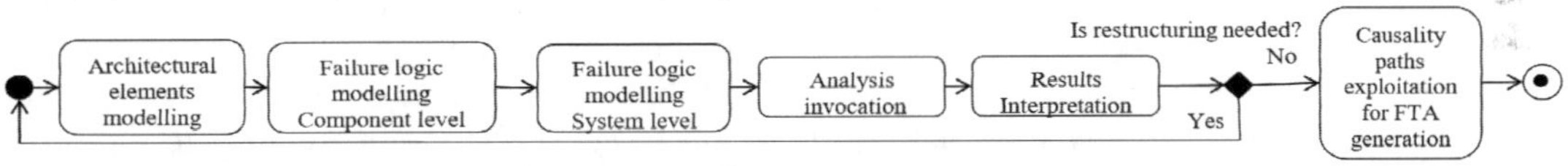

Figure 1. Approach overview, adapted from [4].

4. Modelling

In this section, first, we explain the system level assumptions, then, we model the ACS system (textually described in Section 2).

4.1. *System level assumptions*

Assumption 1. The system starts computing torques when valid estimates of the input sensor readings are available. An external component checks the estimated values of the input sensors and, in response to the valid estimates, activates the ACS system. This component is assumed to be analyzed already as per ECSS standards [6][7] and is out of the scope of this paper. The validity check in the component determines that the sensors are activated and registering their perceptions, and does not reason on the value itself being registered. Therefore, this assumption rules out the following failure types/modes on the input ports (1) Omission, (2) Commission, (3) Early, and (4) Late.

Assumption 2. Each of the components has been internally verified as per ECSS standard verification guidelines. Thus, none of the components in the system could act as the source of the failure. Rather, a component may only propagate, transform or sink failures.

IOP Conf. Series: Materials Science and Engineering **351** (2018) 012001 doi:10.1088/1757-899X/351/1/012001

4.2. Modelling of ACS in SafeConcert

The ACS described in Section 2 is modeled using CHESS-ML/SafeConcert, which enables for representing the system architecture utilizing the component based paradigm. ACS itself is represented as a composite component, which is composed of all the other components, collectively defining the function of the system. Figure 2 illustrates the component based design of the software system with injected failures on the input ports. ACSComposite component has three inputs and four output ports providing the interfaced sensors readings and actuators commands respectively. More specifically, the three input ports sunEstVec, angVelocity and gyroDistTorques represent Sun direction, angular velocity of the satellite and measured disturbance torques, respectively. The output ports ctrlTorque, propTorque, derTorque and feedforwardTorque corresponds to the control torque selected to be applied to the satellite for achieving the desired orientation, the torque computed through proportional controller, the torque computed through derivative controller and compensating cross coupling torque respectively.

According to the assumptions listed in Section 4.1, on the input ports, value failure could be experienced. As shown in figure 2, this value failure, which is of two types i.e. valueSubtle and valueCoarse, is injected on each input port by using the appropriate CHESS stereotype, called *FPTCSpecification*. Moreover, the failure behavior of each component is given by using the *FLABehavior* stereotype. This behavior is described as follows.

PDController component propagates and sinks the failures exposed on its input ports. If there is a valueCoarse failure on any of the input ports, the produced torques will also have a valueCoarse failure. The valueSubtle failure on sunEstVec input port is propagated to the output ports. Whereas, a valueSubtle failure occurs on the angVelocity input port, the component detects it via a low pass filter and sinks it and produces noFailure on the output ports. Similarly, FeedforwController component also detects a valueSubtle failure on its input port using a low pass filter and sinks it and produces no failure on output port. FeedforwController, however, propagates the valueCoarse failure to FeedforwardTorque output port. SteerController component propagates both the valueSubtle and valueCoarse failure on the sunestVec input port to the output torque. Finally, the TorqueSelector component also propagates both the valueSubtle and valueCoarse failure on the input ports to the ctrlTorque output port.

5. Analysis of ACS via ConcertoFLA

To apply ConcertoFLA, all the three input ports of the system shall be injected with possible failures. Considering the assumptions in Section 4.1, the possible failures are limited to only value failures i.e. valueSubtle and/or valueCoarse. This suggests two possible failures for each port. However, in this paper, for space reasons, we only consider one input port i.e., sunEstVec. This port receives faulty data. To compute the torques, PDController and SteerController rely on the data received on sunEstVec. Thus, the computed torques are also faulty (valueCoarse or valueSubtle failure). The control torque selected by TorqueSelector, based on the context i.e., satellite current attitude opposite to pointing object or closer to the pointing object (see Section 2), will also be faulty. The resulting torque, in case of valueSubtle failure on the output port, when applied on the satellite, could result in loss of mission. The valueCoarse failure could result in major mission degradation due to not pointing accurately to the object of interest. Figure 3 shows the propagated failures on the output ports.

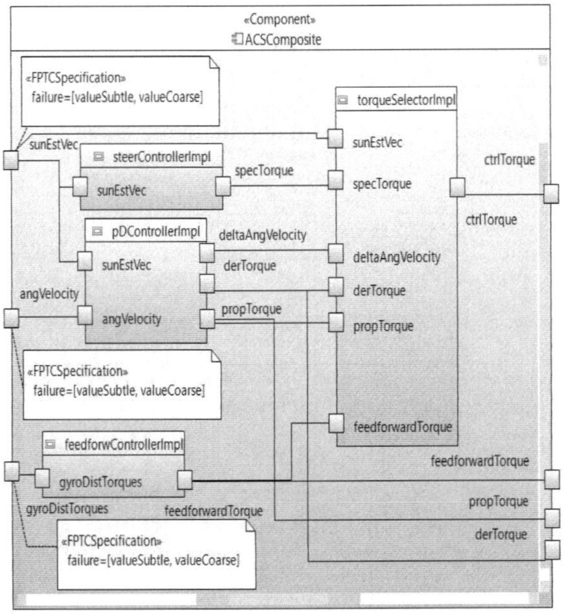

Figure 2. ACS system before the analysis.

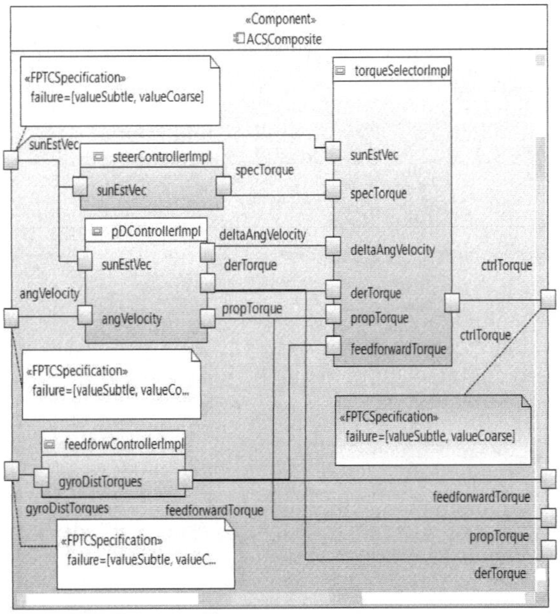

Figure 3. ACS system with analysis results.

6. ECSS-compliant FTA via ConcertoFLA

ConcertoFLA also calculates the failure propagation paths and produces their representation according to the specifications of FlaMM meta model (see [14] for FlaMM structure and corresponding XML Schema). Figure 4 shows some XML snippets representing the failure propagation paths. The yellow highlighted text illustrates the failure behavior on the CtrlTorque output port along with the information on previous failures. Figure 5 displays the same failure propagation for ACS composite component in a tree-like view. The component exhibits valueCoarse failure on the ctrlTorque output port. This is a system level failure and could be described as a top event of the FTA. The valueCoarse failure could be collapsed in the tree and the contributing previous failures could be traced, as depicted in figure 5. The ports exhibiting previous failures could be used as the contributing events of top level event, in the FTA. The information regarding the owning component of the port could also be retrieved from the tree, to uniquely identify ports with same label e.g., in the ACS case. This also conforms to one of the requirement from IEC 61025, which specifies that the FTA events shall be uniquely identified. Moreover, all the listed components in the tree could also be collapsed further to view the failure behavior displayed as tree nodes. This information is crucial to identify the relationship between an event and its descendants, otherwise depicted using logical gate symbols in FTA. Thus, the FlaMM model contains all the required information to construct an ECSS conformant FTA. Figure 6 illustrates ECSS-compliant partial FTA manually produced from the results stored in FlaMM model upon a successful execution of ConcertoFLA analysis. For the sake of clarity, it should be noted that given the qualitative nature of the ConcertoFLA analysis the fault tree cannot contain quantitative information.

7. Related work

To develop dependable systems many researchers are utilizing model based methods. In [9] and in [4], some of these works were already discussed (focus on dependability modelling and analysis). Thus, in this paper, we limit our attention to those works that have considered automatic generation of analysis artifacts based on requirements coming from standards. To enhance the V-model process of the avionics product development, in [15], authors propose a model based tool chain which embraces the end-to-end process covering all phases of software development lifecycle and focusing explicitly on dependability aspects. They also use a SySML-metamodel extension and consider FTA analysis. However, the integration of dependability analysis is still in progress and no generation is mentioned.

In [16], a model based approach is proposed to augment the structure and behavior models of railway domain system with fault related information. Effects of the presence of faults are analyzed in system and are used to complement the preliminary Failure Modes, Effects and Analysis (FMEA).

Figure 4. XML based ACS FlaMM model.

Figure 5. Tree-like ACS FlaMM model.

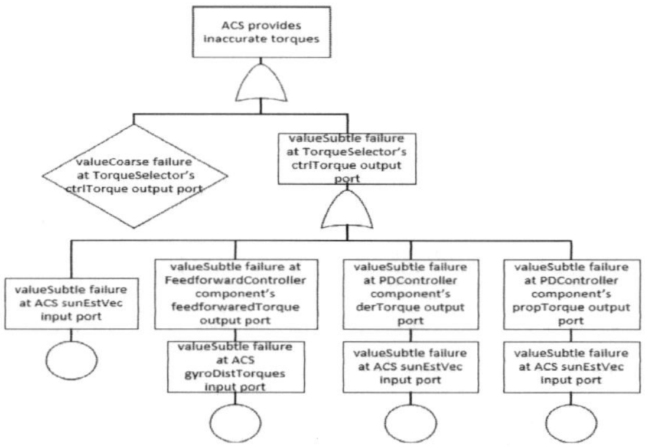

Figure 6. ECSS-compliant Partial FTA of ACS generated manually.

8. Conclusion and future work

In this paper, we have explored the usage of the tool-supported CHESS-methodology in the context of Attitude Control Systems Engineering. More specifically, we have used ConcertoFLA, the dependability analysis technique incorporated within the CHESS methodology. Via ConcertoFLA, we have modeled an Attitude Control System, its failure behavior, and performed analysis. Based on the analysis results, we could elaborate on FTA-results generation in the context of ECSS standards and more generally in the context of IEC 61025.

In the future, we aim at implementing a plugin to support a proper generation of FTA, in compliance with the largely accepted graphical concrete syntax. Based on ongoing work in the context of AMASS [17] and previous work [18] [19], we also aim at generating product and process-based argument fragments. This generation is also expected to connect ECSS-compliant product artifacts with ECSS-compliant process artifacts, since dependability analysis results constitute expected work products in a software development process model.

9. References

[1] Fortescue P and Stark J 1998 *Spacecraft Systems Engineering* (John Wiley & Sons ltd West Sussex England)

[2] Markley F L and Crassidis J L 2014 *Fundamentals of Spacecraft Attitude Determination and Control* (Springer New York/Heidelberg/Dordrecht/London)

[3] PolarSys CHESS, https://www.polarsys.org/chess/ (last accessed 03/08/2017)

[4] Gallina B, Sefer E and Refsdal A 2014 Towards safety risk assessment of socio-technical systems via failure logic Analysis IEEE Int., Symp., on Software Reliability Engineering Workshops (ISSREW) Naples Italy pp 287-292
[5] European Cooperation for Space Standardization ECSS-Q-ST-30C Space product assurance - Dependability 06/03/2009
[6] European Cooperation for Space Standardization ECSS-Q-ST-40-12C Space product assurance - Fault tree analysis – Adoption notice ECSS/IEC 61025 06/03/2009
[7] Internationl Electrotechnical Commission IEC 61025 fault tree analysis (FTA) IEC 61205:2008
[8] U.S. Nuclear Regulatory Commission NUREG 0492 1991 Fault Tree Handbook
[9] Montecchi L and Gallina B SafeConcert: a metamodel for a concerted safety modeling of socio-technical systems 5th Int., Symp., on Model-Based Safety and Assessment (IMBSA) Trento Italy September 2017
[10] SysML v1.4 Specification Release September 2015 http://www.omgsysml.org/specifications.htm
[11] Gallina B, Javed M A, Muram F and Punnekkat S 2012 Model-driven dependability analysis method for component-based architectures IEEE Proc., of the Euromicro Software Engineering and Advanced Applications (SEAA) Conf., Cesme Izmir Turkey
[12] Grunske L and Han J 2008 A comparative study into architecture-based safety evaluation methodologies using AADL's error annex and failure propagation models 11th IEEE High Assurance Systems Engineering (HASE) Symp., Nanjing China pp 283–292
[13] Wallace M 2005 *Modular architectural representation and analysis of fault propagation and transformation* Electronic Notes in Theoretical Computer Science vol, 141 issue 3 pp 53–71
[14] CONCERTO Deliverable D3.3 November 2015 Design and implementation of analysis methods for nonfunctional properties – Final version
[15] Hovsepyan A, Landuyt D V, Beeck S, Michiels S, Joosen W, Rangel G, Briones J F and Depauw J 2014 Model-driven software development of safety-critical avionics systems: an experience report, in Proc., Int., Workshop on Model-Driven Development Processes and Practices (MD^2P^2) CEUR Workshop Proceedings pp 28–37
[16] Olmo J, Poza J, Garramiola F, Nieva T and Aldasoro L 2017 Model driven hardware-in-the-loop fault analysis of railway traction systems IEEE Int., Workshop of Electronics, Control Measurement, Signals and their Application to Mechatronics (ECMSM) Donostia-San Sebastian pp 1-6
[17] AMASS (Architecture-driven, Multi-concern and Seamless Assurance and Certification of Cyber-Physical Systems) http://www.amass- ecsel.eu
[18] Alajrami S, Gallina B, Sljivo I, Romanovsky A, Isberg P 2016 Towards cloud-based enactment of safety-related processes Proc., of the 35th Int., Conf., on Computer Safety, Reliability and Security (SAFECOMP) Trondheim Norway Lecture Notes in Computer Science vol, 9922 pp 309-321
[19] Gallina B 2014 A model-driven safety certification method for process compliance IEEE Proc., of Int., Symp., on Software Reliability Engineering Workshops (ISSREW) pp 204-209

Acknowledgement
This work has been partially financially supported by the AMASS project [17]. We thank S. Puri and P. Isberg for their valuable explanation regarding CHESS tool-set and ConcertoFLA implementation.

IOP Conf. Series: Materials Science and Engineering **351** (2018) 012002 doi:10.1088/1757-899X/351/1/012002

Steel Bar corrosion monitoring based on encapsulated piezoelectric sensors

Ying Xu, Tianyou Tang

Shenzhen Key Urban & Civil Engineering Disaster Prevention & Reduction Laboratory, Shenzhen Graduate School, Harbin Institute of Technology, Shenzhen 518055, China

cexyx@hotmail.com

Abstract. The durability of reinforced concrete has a great impact on the structural bearing capacity, while the corrosion of steel bars is the main reason for the degradation of structural durability. In this paper, a new type of encapsulated cement based piezoelectric sensor is developed and its working performance is verified. The consistency of the finite element simulation and the experimental results shows the feasibility of monitoring the corrosion of steel bars using encapsulated piezoelectric sensors. The research results show that the corrosion conditions of the steel bars can be determined by the relative amplitude of the measured signal through the encapsulated piezoelectric sensor.

1. Introduction

Steel bars play an important role in strength, durability, prevention of crack propagation and other aspects in a structure. Steel corrosion can cause cracks inside the concrete along the bars, greatly reducing the structural bearing capacity and service life. Therefore, effective monitoring methods are needed for reinforced concrete structures to assess the extent of steel corrosion. The following represent the more mature steel corrosion detection and monitoring methods at present: the half-cell half-potential method, the concrete resistivity method, chloride ion content testing, concrete corrosion cracks sizing and ultrasonic parameters assessment [1]. These methods are indirect measurements, and can only qualitatively describe the degree of corrosion; they cannot achieve quantitative on-line monitoring. For large structures it is even more difficult to achieve all-round monitoring.

The piezoelectric sensor has good linearity, fast response, ease of processing, low cost and other advantages, giving the piezoelectric sensor growing popularity in health monitoring. In order to monitor the early curing strength of concrete pouring, Yang Xiaoming, Li Zongjin et al. [2] developed a new cement-based sensor and researched the relationship between the cement-based piezoelectric sensor's frequency independence and load amplitude with the voltage amplitude. They demonstrated the testability of the cement-based sensor on concrete structures. However, research in corrosion monitoring with piezoelectric sensors is still in the experimental stage, and there is also a lack of corresponding research in the aspect of uniform corrosion of steel bars.

2. Encapsulated piezoelectric performance research

2.1. Encapsulated piezoelectric sensor production process

In this paper, PZT5 piezoelectric ceramics are used. When the piezoelectric sensor and the ceramics work together in concrete, the encapsulating material does not change the performance of the concrete or the sensor. Based on this principle, AB glue was first used to encapsulate the PZT chip, then neat cement, then it was coated with conductive silver glue for an electromagnetic shielding effect. The performance parameters are in Table 1.

Table 1. PZT5 bare chip parameters

Size(mm)	Piezoelec-tric coefficient d33 (pC/N)	Relative dielectric constant	Curie tempera-ture (℃)	density (g/cm)
Φ10×1.0	670	5800	200	7.45

Production process:

(1) The wire is joined. Soldering is used to join the PZT5 bare chip and wire. For the receiving sensor, the aluminum of the shielded wire is twisted together, and then secured to the outside of the sensor, thereby shielding the noise.

(2) The PZT5 bare chip is encapsulated. After evenly mixing the AB glue in a 1: 1 ratio, the exposed PZT copper wire is completely encapsulated to prevent a short circuit of the sensor due to water entering, and after curing, conductive silver paste applied to the surface (see Figure 1-a).

(3) The cement base is produced. The encapsulated PZT pieces are put into a 20 × 20 × 20mm square mold, the water-cement is added in a ratio of 0.5 of the cement paste, and cured for a period of time (see Figure 1-b).

(4) Shielding is carried out. Conductive silver paste is applied to the outer surface of the cured cement base, and the shielded wire is glued to the sensor surface, Figures 1-c show the receiving end of the piezoelectric sensor after shielding.

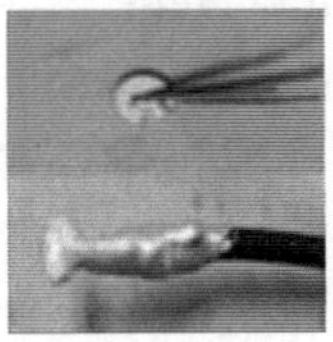

a) AB glue encapsulation b) cement base production c) shielding processing

Figure 1. Fabrication process of encapsulated piezoelectric sensor

2.2. Encapsulated piezoelectric sensor shielding performance

The encapsulated piezoelectric sensor and a piezoelectric sensor without shielding processing are directly connected to the oscilloscope, and a signal can be obtained as shown in Figure 2. It can be seen that after shielding the noise signal basically disappears. At the same time, after FFT conversion it can be seen that the noise signal interference frequency is about 50 Hz, and the peak is about 1.5 mV.

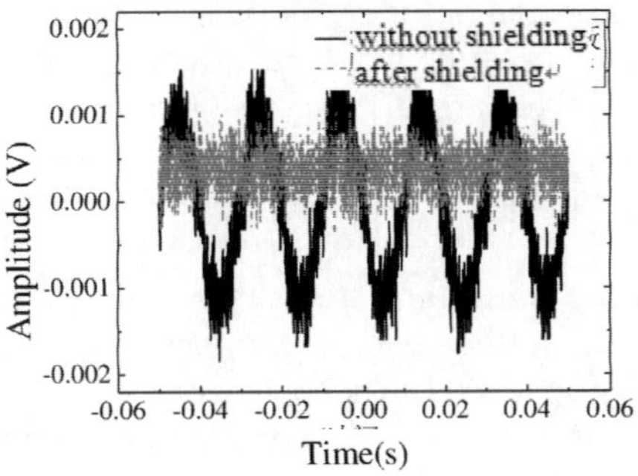

Figure 2. Environmental noise signal

2.3. Encapsulated piezoelectric sensor stability
The encapsulated piezoelectric sensor is used for continuous monitoring of the signal changes during the hydration process of the neat cement mortar, and the signal and waveform is read every 30 minutes. The monitoring process is shown in Figure 3.
The average values of two simultaneous test groups are taken, and the results are basically similar. Due to limited space, only the average is shown as an example. The results are shown in Figure 4.

Figure 3. Gravel and cement hydration

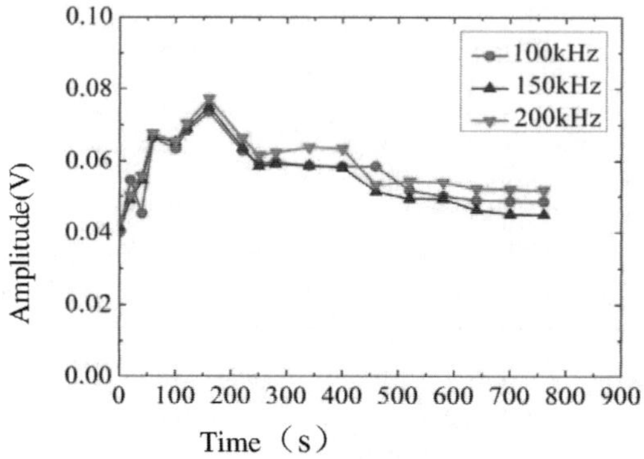

Figure 4. Mortar amplitude time curve

When the ultrasonic signal from the encapsulated piezoelectric sensor propagates in the neat cement mortar, it demonstrates stability after the first increase. At about two and a half hours there is a greater decline in signal amplitude, and the signal tends to stabilize after 10 hours. This is due to the fact that the strength has increased after cement hydration, resulting in limited deformation of the sensor, so that the sensitivity of the sensor to the signal falls [3]. These results indicate that the encapsulated piezoelectric sensor can maintain its stability in the cement mortar after standing for 10 hours.

3. Steel corrosion monitoring experiment

3.1. Reinforced cement mortar specimen production
A number of steel bar - cement mortar specimens were produced, their mixture ratios and size are shown in Table 2. Ordinary Portland cement of grade 32.5, sand and ribbed steel bars with a diameter of 10 mm and length of 140 mm were used. The three parallel identical specimens produced were monitored. The completed specimens are shown in Figure 5.

3.2. Steel rebar corrosion monitoring
The pre-made steel corrosion specimen was placed in a 5% concentration NaCl solution and soaked for 48 hours with the solution making full contact with the steel to enable corrosion to occur. Then the specimen was connected with the DC power source. The DC power source current was set to 0.1 A, and the data was measured at every 2% of the theoretical corrosion rate, i.e. the data was read once every 15.5 hours. The monitoring continued for 155 hours. For each working condition a sinusoidal excitation signal was used, the voltage amplitude was plus or minus 5 V. Each frequency screening was performed, preferably at a large amplitude band, due to the fact that the high frequency amplitude of these piezoelectric ceramics is rather large. Thus, this monitoring used the ultrasonic frequency band of 180-200 kHz for continuous testing which is commonly applied in non-destructive testing in the structural health monitoring field. Figure 6 shows the experimental setup for uniform corrosion monitoring of reinforced bar.
The tightness of the attachment of the sensor and steel rebar will affect the sensitivity of the sensor to the received signal. This can cause a system error in the measurement values of the experiment and make the subsequent horizontal comparison difficult. In order to eliminate this error, the relative amplitude (ξ) as defined herein is the ratio of the corresponding monitor signal amplitude of any corrosion rate (A_T) and the monitor signal amplitude of the corrosion rate at time 0 (A_0), namely:

$$\xi = \frac{A_T}{A_0} \quad (1)$$

Table 2. Sample size

Mix Ratio			Specimen size	Compressive strength (MPa)
Cement	Sand	Water		
1	2	0.5	180×40×40	26.97

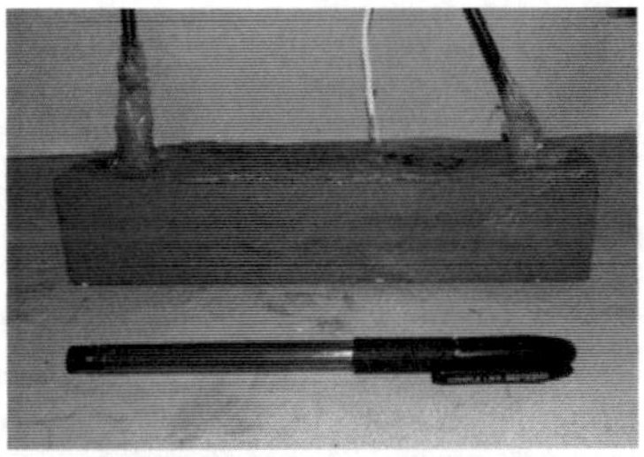

Figure 5. Uniform corrosion specimen

Figure 6. Uniform corrosion monitoring of reinforced bar

Corrosion monitoring was performed for the three parallel identical specimens, and the signals collected underwent bandpass filtering noise reduction to extract the amplitude of each frequency monitoring signal. The curve of each self-monitoring signal amplitude over power-on time was obtained, the relative amplitude ξ defined using formula (1) was used, and finally the curve of relative amplitude over power-on time was obtained. After obtaining the respective relative amplitude - corrosion rate curves, the three specimens were averaged, and the final average relative amplitude – power-on time curves and average relative amplitude - corrosion rate curves were obtained. These are shown in figures 7 and 8.

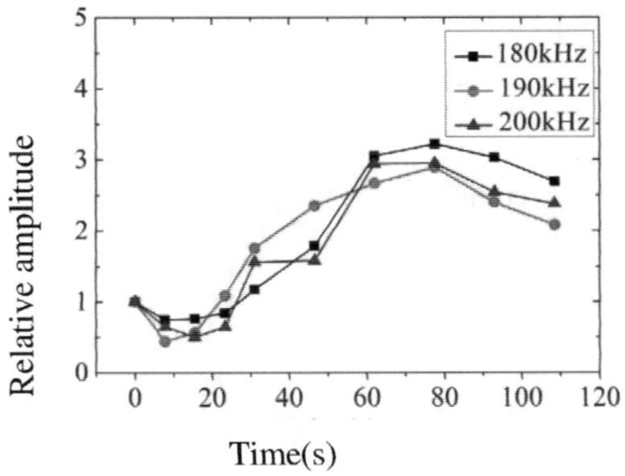

Figure 7. Relative amplitude-time curve

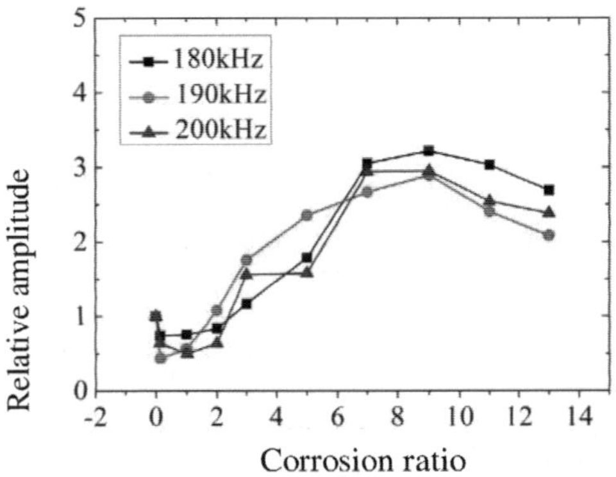

Figure 8. Relative amplitude-corrosion curve

Figures 7 and 8 are the relative magnitude - time curve and relative amplitude - corrosion rate curve, respectively. It can be seen from the figures that:

(1) When the power-on time is less than 20 hours and the actual corrosion rate is at 2% or less, the relative amplitude is between 0 and 1, and there is a decrease. When the steel begins to corrode, the steel surface exhibits localized pitting, and the signal amplitude is reduced when reaching the terminal due to defects in the surface causing signal emission passing through the steel bar [4,5]. Another possible reason is that before the specimen starts cracking, rust expansion increases the adhesive force

between the steel and concrete, so that the adhesive force between steel and concrete does not drop and actually slightly increases [6].

(2) When the power-on time is between 20 and 80 hours, the actual corrosion rate is 2% to 7%, the relative amplitude is between 1-3, and the relative amplitude greatly increases as the corrosion rate increases. This is because the accumulation of rust in the corrosion process can cause the effective diameter of the steel rebar to decrease, and form a corrosion layer at the rebar and mortar interface, causing the adhesive force to gradually drop [5]. After the adhesive force between the rebar and mortar drops, debonding occurs in both, the original signal transmitted from the rebar surroundings to the mortar is gradually reduced due to the decline of the degree of adhesion in both, and the signal propagated from the steel is thereby increased3[7]. The relative signal amplitude reaches the maximum value, marking the point where the steel and mortar are completely debonded, and the formation of a complete layer of rust.

(3) When the actual corrosion rate is between 7-13%, the relative amplitude is between 2 and 3, and the relative amplitude tends to be stable and decreases slightly as the corrosion rate increases. It may due to that with the further corrosion of the steel, the mortar surrounding the steel rebar encounters greater expansion force and cracking, resulting in attenuation in the high-frequency monitoring signal due to the crack damage, and with the deepening and extension of the cracks the amplitude decreases. Therefore, the corrosion rate can be determined according to the relative magnitude and change trends at certain times.

4. Conclusions

(1)The encapsulated piezoelectric sensor performance meets the corrosion monitoring requirements. The shielded encapsulated sensor can shield the surrounding noise signals with a frequency of 50 Hz, and amplitude of about 1.5 mV. After cement-base coating it is able to reduce each frequency signal amplitude, and can attenuate signal amplitude in the water by about 94.8%, while as the size of the cement base increases, the amplitude attenuation also gradually increases.

(2) From the experimental results, it can be seen that the experimentally measured relative amplitude – corrosion rate curve and the simulated curve show a generally consistent tendency, but there is still a certain discrepancy. When the corrosion rate is less than 7%, the difference between both is rather small. When the corrosion rate is 0-2%, the relative amplitude decreases; when the corrosion rate is 2-7%, the relative amplitude increases along with the corrosion rate. When the corrosion rate is greater than 7%. This corrosion rate satisfies the requirements of structural safety monitoring in actual projects.

5. References

[1] LIU Ying, WANG Bin. Talking about steel corrosion in concrete and its detection methods.[J].Northern Communications, 2012(7):61-62.

[2] YANG xiaoming,LI Zongjin.Performance of a Novel Cement-Based Piezoelectric Sensor[J].Chinese Journal of Sensors And Actuators,2012,25(3): 351-354.

[3] SUN Wei,YUAN Yanshi,JIANG Shaofei. The Influence of Concrete Boundedness On Stress Wave Propagating and its correction[J]. China Concrete And Cement Products, 2014(4):55-57.

[4] Tri H Miller, Tribikram Kundu, Jiaqiang Huang, et al. A New Guided Wave–Based Technique for Corrosion Monitoring In Reinforced Concrete[J]. Structural Health Monitoring, 2012,12(1): 35–47.

[5] Samir Mustapha, Ye Lu, Jianchun Li, et al, Damage Detection In Rebar-Reinforced Concrete Beams Based on Time Reversal of Guided Waves[J]. Structural Health Monitoring,2014, 13(4): 347–358.

[6] LI Bin,HE Xianfeng,LUO Zhimin. Effect of Corrosion On Cohesive Force of Reinforced Concrete[J]. Industrial Construction,2001,31(5): 12-14.

[7] ZHONG Qianqian. Study of Ultrasound Monitoring Reinforcing Steel Corrosion Based on Piezoelectric Sensor[D]. Ji nan:Dissertation of University of Jinan,2012: 8-13.

Acknowledgement

This work is supported by funding of NSFC (51778191) and basic project of Shenzhen Science & Technology Program (JCYJ20170307145853300).

IOP Conf. Series: Materials Science and Engineering **351** (2018) 012003 doi:10.1088/1757-899X/351/1/012003

Effect of tooth profile modification on wear in internal gears

M S TUNALIOGLU[1*], B TUC[2]

[1] Hitit University, Department of Mechanical Engineering, Corum, TURKEY
[2] Başkent University, Department of Mechanical Engineering, Ankara, TURKEY

*mstunalioglu@gmail.com

Abstract. Internal gears are often used in the automotive industry when two gears are required to rotate in the same direction. Tooth shapes, slippage speeds at the beginning and end of meshing are different according to the external gears. Manufacturing of internal gears is more difficult than external gears. Thus, it is necessary to determine the working conditions and wear behavior of internal gears carefully. The profile modification method in terms of strength and surface tension of the gear mechanism are performed in order to increase the load-carrying capability. In this study, profile modification method was performed in the internal gears to reduce the wear on the teeth. For this purpose, the wear of the internal gears was theoretically investigated by adapting the Archard wear equation to the internal gears. Closed circuit power circulation system was designed and manufactured to experimentally investigate the wear in internal gears. With this system, wear tests of gears made of St 50 material without profile modification and different profile modifications were made and the results were compared. Experimental study was performed in the same loading and cycle time conditions to validate the theoretical results and it was seen that the results are compatible. According to the experimental results, it is seen that in the internal gears, when profile modification done the wear is decreased in the teeth tip region.

1. Introduction

Internal gears are different from the external gears in that their teeth are turned inward rather than outside the gear center. Internal gears having concave tooth profile mate with external gears having convex tooth profile which provides some advantages; low contact stress, low sliding velocity and high gear ratio [1]. Internal gears are widely used in automotive and aerospace industries, differential boxes and cranes as external sun gears of planetary mechanisms due to their compact structure, large torque-to-weight ratio, high gear ratio, reduced noise and vibration, etc [2].

Tooth profile modification; to increase the load carrying capacity of the gear mechanisms in terms of strength and surface pressure, to prevent the undercutting, to set the gear mechanisms between the gear axle offset, to reduce the slippage between the gears and accordingly to reduce the wear [3]. Several theoretical and experimental studies have been carried out on internal gears. In studies carried out to the geometry of internal gears; the effect of the rim thickness on the root stresses was investigated experimentally and theoretically [4-6]. These studies provide the manufacturing of internal gears with optimum rim thicknesses. In studies to examine the stresses in the tooth root of internal gears; effect of tensile and compression stresses in gear root examined [7-9]. Analytical formulations, finite element simulations, strain-gauge and photoelasticity based experimental procedures were used in stress measurements in these studies. Several studies have been carried out in order to prevent wear in internal gears [10-11]. In these studies, internal gears were covered with coating material to prevent wear. In this study, tooth profile modification method was used in the gears to reduce wear on the internal gears. The wear occurred in the gears without profile modification and a different profile modification under the same conditions was compared in theoretically and experimentally.

2. Wear model in internal gears

The teeth of the gears, which are in contact with each other, make sliding and rolling. As a consequence of these movements, wear occurs on the surfaces of gears. In the Archard wear equation[12];

$$\frac{V}{s} = K\frac{W}{H} \tag{1}$$

(V) is volume of the abraded material, (s) is slippage distance between two surfaces, (K) is the friction coefficient, (W) is applied load and (H) is surface hardness of the material. In order to use this equation in gears, it is necessary to take into consideration the sliding and rolling which show differences from the beginning to the end of the contact of the teeth. Flodin [13] adapted this equation to external gears.

$$h_p = \int_0^s kP\,ds \tag{2}$$

Here; (h_p) is the wear depth in point (p), (s) is the sliding distance between two contacting surfaces, (k) is the wear coefficient and (P) is the regional contact pressure. To contacting gear pairs, namely expressing wear of any contacting point of tooth profiles of pinion-internal gear during coupling depending on rotation cycle count, wear depth at a point (p) after (n) cycle count can be expressed;

$$h_{p,n} = h_{p,(n-1)} + kP_{p,(n-1)}s_p \tag{3}$$

where ($h_{p,n-1}$) is the wear depth of same point in the previous cycle, ($P_{p,n-1}$) is the pressure at point (p) in the previous cycle and (s_p) is the sliding distance of point (p). In order to determine the slip distance at the contact points, examine the two points, p_1 and p_2, which are opposite to each other, from the teeth of the pinion and the internal gear through the coupling. These points were investigated in three different positions during the coupling. In the first position, p_1 and p_2 coincide in the beginning of contact and there is no difference in distance between them. In the second position; when p_1 is exiting coupling, p_2 is still in contact region. Thus, there is a distance of s_{p1} between p_1 and p_2. In the third position; when p_2 is exiting coupling, there is a distance of s_{p2} between p_1 and p_2. Point p_2 moves a distance of $2a_H.(U_2/U_1)$ when p_1 moves a distance of $2a_H$ (Hertz contact length) along the contact length in the contacting teeth pair where U_1 and U_2 are peripheral speeds of pinion and internal gear along the contact line, respectively. Similarly, p_1 moves a distance of $2a_H.(U_1/U_2)$ when p_2 moves a distance of 2a along the contact length in the contacting teeth pair.

$$s_{p1} = 2a_H\left(\frac{U_1 - U_2}{U_1}\right) \quad ; \quad s_{p2} = 2a_H\left(\frac{U_2 - U_1}{U_2}\right) \tag{4,5}$$

If equations 4 and 5 are written in Equation 3, wear depth equations for pinion and internal gear are respectively;

$$h_{p,n} = h_{p,(n-1)} + kP_{p,(n-1)}^{2} a_{H} \left(1 - \frac{U_2}{U_1}\right) \; ; \; h_{p,n} = h_{p,(n-1)} + kP_{p,(n-1)}^{2} a_{H} \left(1 - \frac{U_1}{U_2}\right) \quad (6,7)$$

Pinion and internal gears peripheral speeds (U_1,U_2), Hertz pressure on the contact points of the teeth profiles (P) and Hertz contact area of tooth profiles (a_H) described in the previous work [10].

3. Experimental study

3.1. Test gears

The teeth numbers of pinion and internal gears used in the tests were 17 and 75 respectively and they were manufactured from St50 steel. The surface hardnesses of the gears were 160-170 HB. In the experiments, gears without profile modification, 0.095, 0.195 and 0.410 profile modification gears were used.

3.2. Test equipment

A pinion-internal gear fatigue and wear testing apparatus having the same working principle with the FZG [13-16] closed circuit power system was manufactured to perform wear tests (Figure 1). The apparatus which allows investigation of wear in various load and speed conditions seen in Figure 1 consists of two gear boxes having same gear ratio. There are control panels in the test setup where the temperature control and the number of revolutions of the driven motor can be adjusted. The system is loaded with 100 Nm and the motor speed is constant at 1500 rpm.

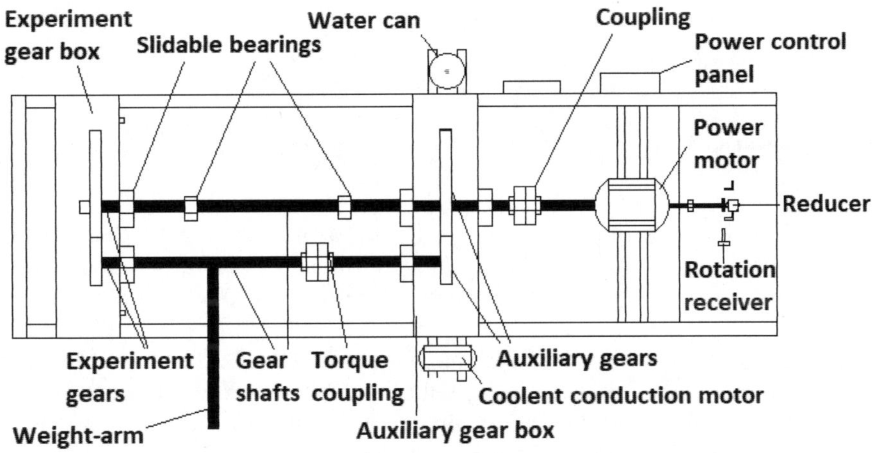

Figure1. Schematic drawing of testing apparatus

In the experiments, immersion type lubrication system was used. The oil temperature is kept constant at 23 ± 2 ° C due to the heater-cooler system.

4. Theoretical and experimental results

In order to investigate wear in internal gears, Archards' wear formulation was adapted to internal gears and wear behaviour in different tooth profile modification was determined in the theoretical part of the study with a MATLAB® code written for the obtained formulation. A fatigue and wear testing apparatus having similar working principle with FZG closed circuit power circulation system was designed and manufactured to experimentally investigate the wear in internal gears. the experimental wear depth values were measured with the help of CMM (Coordinate Measuring Machines) [10].

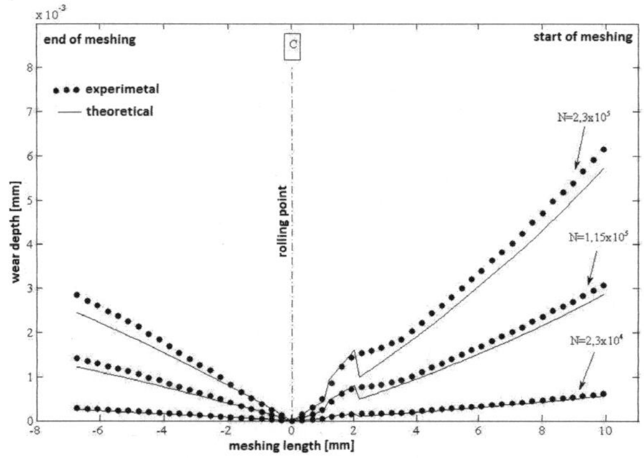

(a) Without profile modification

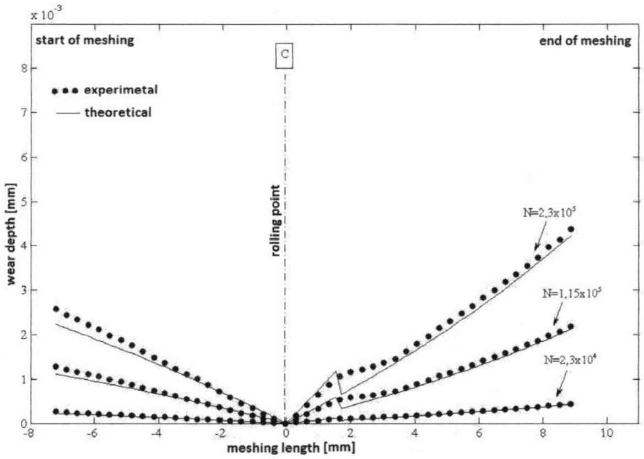

(b) $x_1=x_2=0.095$ tooth profile modification

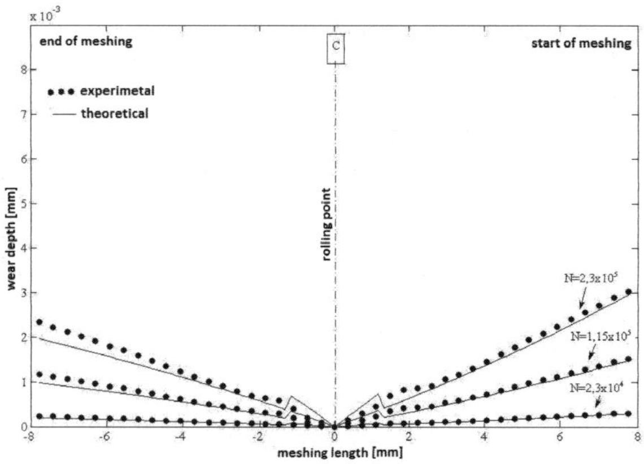

(c) $x_1=x_2=0.195$ tooth profile modification

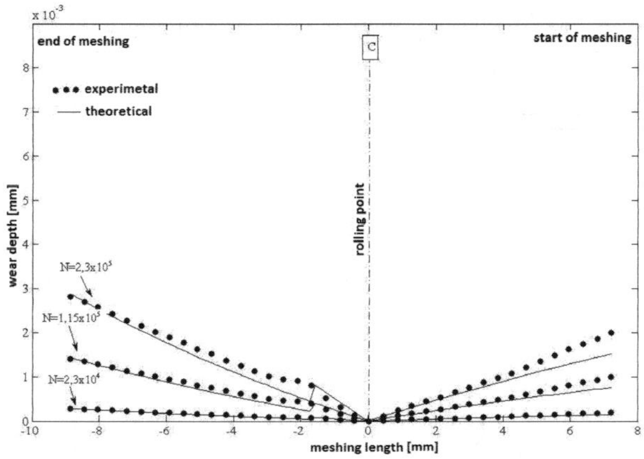

(d) $x_1=x_2=0.410$ tooth profile modification

Figure 2. Variation of wear depth of without profile modification and various profile modification internal gears with 1500 rpm motor speed and 100 Nm torque

According to both the theoretical and the experimental results (Fig.2), the maximum wear is seen in tooth tip region where the internal gear begins to coupling. When Figure 2 is examined, it is seen that experimental and theoretical results are compatible. The equation, which is theoretically used to determine the depth of wear occurring in the teeth, can be used with 2-3 % error when the difficulties of the manufacturing of internal gears, the duration of wear tests and working conditions are considered. When the wear depths of the internal gears without profile modification and different profile modification are theoretically and experimentally compared (Fig. 2 a-d); the wear depth values decrease by 18.4 %, 27.5 % and 39.2 % respectively, when the profile modification amount increases. When the amount of profile modification is increased in the internal gear, wear is reduced in the tooth tip region of the internal gears. However, when the amount of profile modification exceeds a certain amount, the length of entry of meshing reduced, thus the load will more effect the tooth root region and increase the wear depth (Fig. 2 a & d).

5. Conclusion

In this study, the tooth profile modification method was used to reduce the wear on the internal gears. Theoretical and experimental studies have been carried out to investigate the effect of the amount of profile modification on wear. Theoretical results are solved by using Matlab program and applied to internal gears under different profile modification conditions. Experimental studies have been carried

IOP Conf. Series: Materials Science and Engineering **351** (2018) 012003 doi:10.1088/1757-899X/351/1/012003

out on the same conditions to confirm theoretical studies. As a result of this study, the following conclusions were reached.

1. The theoretical equation, which is adapted to internal gears and used with 2-3 % error in different speed and torque conditions, can be used for different profile modification amounts to measure the depth of wear of the internal gears.
2. The profile modification method in the internal gears is a method of increasing the strength of the teeth. Wear can be reduced by using the profile modification method in the internal gears. In internal gears when the amount of profile modification increases, the length of entry of meshing increases and the length of the exit of meshing decreases. In other words, the strength of the tooth tip region of the internal gear increases while the strength of the tooth root region decreases. For this reason, when profile modification is performed in the internal gears undercut of the teeth is take into consideration.

6. References

[1] Terauchi Y, Nagamura K, Ikejo K, 1991, Study on friction loss of internal gear drives, JSME International Journal Series III, 34, 106-113.
[2] Chen Z, Shao Y, 2013, Mesh stiffness of an internal spur gear pair with ring gear rim deformation, Mechanism and Machine Theory, 69, 1-12.
[3] Tunalioglu M S, 2011, A research of tooth profile damages in internal gears, Gazi University Institute of Science and Technology, Ph. D. Thesis, Ankara, 24-34.
[4] Karpat F, Engin B, Dogan O, Yuce C, Yilmaz T G, 2014, Effect of rim thickness on tooth root stress and mesh stiffness of internal gears, International Mechanical Engineering Congress and Exposition, 11.
[5] Oda S, Miyachika K, Araki K, 1984, Effects of rim thickness on root stress and bending fatigue strength of internal gear tooth, Bulletin of JSME, 27, 1759-1764.
[6] Chong T H, Kubo A, 1985, Simple stress formulae for a thin-rimmed spur gear. Part 3: Examination of the calculation method and stress state of internal spur gear, Journal of Mechanisms Transmissions and Automation in Design, 107, 418-423.
[7] Sanchez M B, Pleguezuelos M, Pedrero J I , 2007, Calculation of tooth bending strength and surface durability of internal spur gear drives, Mechanisms and Machine Theory, 95, 102-113.
[8] Yang S, 2007, Study on an internal gear with asymmetric involute teeth, Mechanisms and Machine Theory, 42, 977-994.
[9] Ge N, Zhang J, 2008, Finite element analysis of internal gear in high-speed planetary gear units, Transactions of Tianjin University, 14, 11-15.
[10] Tunalioglu M S, Tuc B, 2014, Theoretical and experimental investigation of wear in internal gears, Wear, 309, 208-215.
[11] Tunalioglu M S, Tuc B, Erdin M E, 2017, Effect of coating materials on wear in internal Gears, IJE TRANSACTIONS B: Applications, 30,1468-1477.
[12] Archard J F, 1953, Contact of rubbing flat surfaces, Journal of Applied Physics, 24, 981-988.
[13] Flodin A, Andersson S, 1997, Simulation of mild wear in spur gears, Wear, 207, 16-23.
[14] Flodin A, 2000, Wear of spur and helical gears, Royal Institute of Technology, Doctoral Thesis, Stockholm.
[15] Hargreaves D J, Planitz A, 2009, Assessing the energy efficiency of gear oils via the FZG test machine, Tribology International, 42, 918-925.
[16] Brandao J A, Martins R, Seabra J H O, Castro M J D, 2014, Calculation of gear tooth flank surface wear during an FZG micropitting test, Wear, 311, 31-39.

IOP Conf. Series: Materials Science and Engineering **351** (2018) 012004 doi:10.1088/1757-899X/351/1/012004

A study on the real-time reliability of on-board equipment of train control system

Yong Zhang, Shiwei Li

School of Electronic and Information Engineering, Beijing Jiaotong University, 3 Shangyuancun, Xizhimenwai, Beijing, 100044, China

zhangy@bjtu.edu.cn

Abstract. Real-time reliability evaluation is conducive to establishing a condition based maintenance system for the purpose of guaranteeing continuous train operation. According to the inherent characteristics of the on-board equipment, the connotation of reliability evaluation of on-board equipment is defined and the evaluation index of real-time reliability is provided in this paper. From the perspective of methodology and practical application, the real-time reliability of the on-board equipment is discussed in detail, and the method of evaluating the real-time reliability of on-board equipment at component level based on Hidden Markov Model (HMM) is proposed. In this method the performance degradation data is used directly to realize the accurate perception of the hidden state transition process of on-board equipment, which can achieve a better description of the real-time reliability of the equipment.

1. Introduction

CTCS-3 (Chinese Train Control System Level 3) has been widely used for 300km/h high speed railways in China to ensure train operation safety and efficiency [1]. Real-time reliability evaluation is conducive to establishing a condition based maintenance system for the purpose of guaranteeing continuous train operation. The demand of real-time reliability has brought about changes to the reliability evaluation because the traditional reliability evaluation methods can only reflect the failure distribution of average characteristics of products of the same type under the same operating conditions and is not applicable for evaluating the reliability in real time for a specific product under specific conditions [2]. Some methods have been proposed for the real-time reliability evaluation including classical analysis, state analysis and intelligent assessment analysis [3].

This paper focuses on the method of evaluating the real-time reliability of the CTCS-3 On-Board Equipment (OBE) based on Hidden Markov Model (HMM). It is organized as follows. In section 2 a real-time reliability index system for the on-board equipment is established; in section 3 the general idea of evaluating the real-time reliability of on-board equipment is introduced by analyzing the structure of on-board equipment and a fault tree model is established; in section 4 the method of predicting the performance degradation trend and the process of calculating the reliability of on-board equipment at the component level is described in detail; in section 5 Balise Transmission Module (BTM), a key component of the on-board equipment, is taken as an example to illustrate the calculation process of real-time reliability ; in section 6 some conclusions are given.

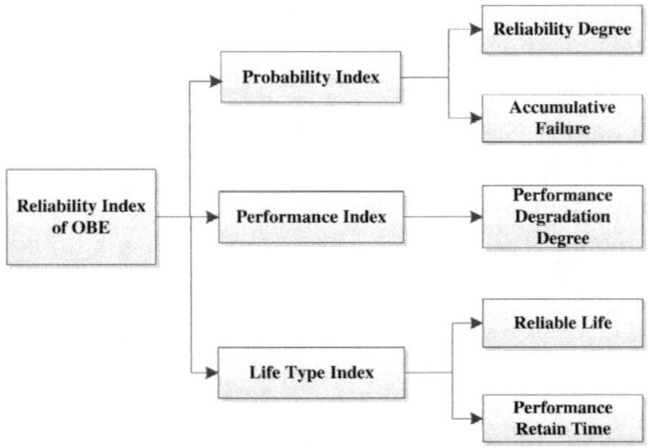

Figure 1. Real-time reliability index system of on-board equipment

2. Establishment of real-time reliability index system

According to the classicial definition of reliability, the real-time reliability is defined as during the service period, the ability of the equipment to perform the required function at current time and under current operation condition. In this paper, we extend the classical reliability index system to set up a reliability evaluation index system for CTCS-3 on-board equipment, as shown in figure 1.

2.1. Probability index

Reliability degree is a quantitive index used to indicate the ablility of the equipment to perform required function under stated conditions for a specified period of time, whichi is denoted as $R(t)$.

$$R(t)=1-F(t)=\int_t^{\infty} g(t)dt \tag{1}$$

Where $F(t)$ is the accumulative failure, used to indicate the probability of failing to perform the required function under stated conditions for a specified period of time, so $F(t)$ is also called the unreliability degree; $g(t)$ is the failure probability density function.

2.2. Performance index

Equipment failure can be divided into two categories: fatal failure and performance degradation failure. The difference between them is that the state of performance degradation failure can't be simply represented by two states (normal/ failure), instead it depends on a number of indicators. Here the performance degradation degree is used to describe the current degradation degree of the on-board equipment, which is defined as

$$K(t)=\begin{cases}0,\left|y(t)-y_0(t)\right|\le\varepsilon\\ 1,\left|y(t)-y_L(t)\right|\le\varepsilon\\ \left|y(t)-y_0(t)\right|/\left|y_L(t)-y_0(t)\right|,other\end{cases} \tag{2}$$

Where $K(t)$ is the performance degradation degree, valued between 0 and 1; $y(t)$ is the measured value of performance parameters; $y_0(t)$ is the optimal value of performance parameters; $y_L(t)$ is the failure threshold of performance parameters.

2.3. Life type index

The reliable life is $t(R)=R^{-1}(R)$, where R is the reliability degree and $R^{-1}(R)$ is the inverse function of R. The performance retaining time, denoted as $T(y_L)$, is the time period during which the performance parameters remain within the specified range

3. Real-time reliability evaluation at system level

3.1. Structure of CTCS-3 on-board equipment

The CTCS-3 on-board equipment is a safety-critical system designed with modular structure and based on fail-safe principle[4]. It is consists of CTCS-3 Control Unit(C3_CU), CTCS-2 Control Unit(C2_CU), BTM, Radio Transmission Module(RTM), Driver Machine Interface(DMI), Speed and Distance Processor (SDP), Speed and Distance Unit(SDU), Track Circuit Receiver(TCR) , Juridical Recorder Unit (JRU) , Train Signalling Gateway(TSG), Safe Transmission Unit-Vehicle(STU-V), Vital Digital I/O (VDX), General Crypt Device(GCD), Compact Antenna Unit(CAU) and so on. Its structure is shown as in figure 2 [1].

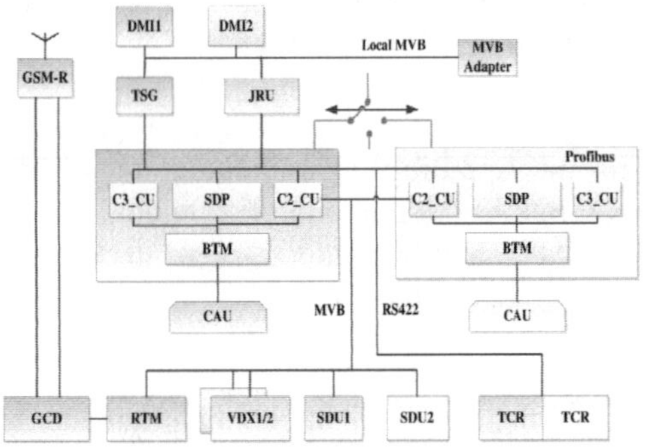

Figure 2. Structure of CTCS-3 OBE.

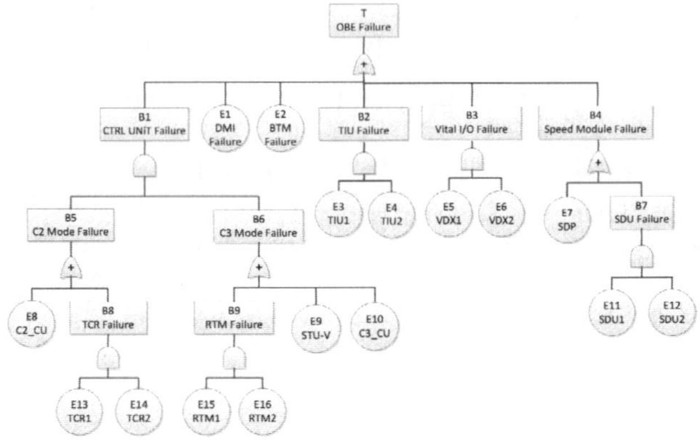

Figure 3. Fault tree model of OBE.

3.2. System-level reliability model

The Fault Tree Analysis (FTA) [5] is adopted to evaluate the system-level reliability. Hot-standby redundancy structure is adopted for the on-board equipment to ensure that the failure of a module can only affect the output of the module itself and will not affect tthe standby module. By analysing the structure and functionality of the on-board equipment, a fault tree model is built, as shown in figure 3. According to the measured performance parameters of the on-board equipment, the performance degradation model can be built to calculate the real-time reliability of the component of the on-board equipment. Finally the system-level reliability can be calculated based on the fault tree model. Therefore the real-time reliability calculation of the on-board equipment is a bottom-up process. The general idea is shown in figure 4.

4. Real-time reliability evaluation at component level

4.1. Matching relation of performance parameters

The Failure Mode and Effect Analysis (FMEA) is a basic reliability analysis method [6]. According to FMEA results and performance parameters related to component failure, a performance index system can be established. Component failure can be divided into two categories: fatal failure and degradation failure, and only the latter one is considered in this paper. The state of a degradation failure depends on a number of parameters, and the performance variation trend of the component is always monotonically increasing or decreasing. When the values of the performance parameters exceed the threshold range, the component will be regarded as in failure state..

The selected performance parameters shall meet the following requirements.

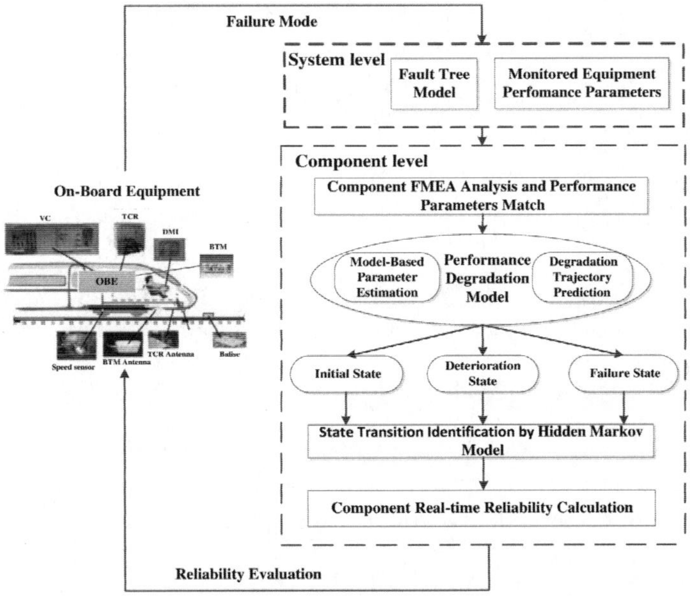

Figure 4. General idea of reliability evaluation of on-board equipment.

(1) They shall be clearly defined and can be measured in real-time;
(2) They shall be able to reflect the failure status of the component;
(3) With the increasing of operation time and worsening of operation condition, the performance parameters shall have significant trend variation.

In order to reasonably select the performance parameters, FMEA is adopted to match the performance parameters with the failure mode of the component. The selection of the real-time reliability parameters normally depends on component structure, expert knowledge and failure log.

4.2. Prediction of performance degradation trajectory

Auto-Regression Integrated Moving Average Model (ARIMA) is a short-term time series forecasting model with high accuracy[7], which can quantitatively describe the process of component performance degradation failure. The modelling steps are as follows.

(1) Series stationarization

Stationeriness test shall be carried out for the collected condition monitoring signal. A non-stationary stochastic series y_{ij} shall satisfy the following equation:

$$y_{ij} = f(t) + g(t) + X(t) \tag{3}$$

Where y_{ij} is the value of *j-th* characteristic value at *i-th* time; $f(t)$ is the non-periodic growth component; $g(t)$ is the periodic component of y_{ij}; $X(t)$ is the component of stationary stochastic processes of y_{ij}. From equation (3) it can be seen that a series can be converted into stationary stochastic series after the non-periodic and periodic trend are separated. Then AR, MA or ARMA can be adopted to build the model.

(2) Model identification and parameter estimation

Through *d-th* order difference, the nonstationary stochastic series $\{y_{ij}\}$ can be converted into a stationary series $\{w_{ij}\}$.Taking ARMA for example, the parameter estimation is as follows.

$$\theta(B)w_{ij} = \phi(B)\varepsilon_{ij} \tag{4}$$

$$E(\varepsilon_{ij})=0, Var(\varepsilon_{ij})=\sigma^2, E(\varepsilon_{mj}\varepsilon_{nj})=0, m\neq n \tag{5}$$

$$\theta(B)=1-\sum_{h=1}^{p}\theta_h B^h, \phi(B)=1-\sum_{h=1}^{q}\phi_h B^h \tag{6}$$

Where $\theta(B)$ and $\phi(B)$ are the autoregressive coefficient polynomial of ARMA and the moving average coefficient polynomial respectively; $\{\varepsilon_{ij}\}$ is a zero mean white noise series. The process of model identification and parameter estimation is firstly to determine the preliminary computational model by analyzing the Auto-Correlation Function (ACF) and the Partial Auto-Correlation Function (PACF), and then to obtain the optimization model according to AIC information norm.

(3) Model prediction

The linear expression of ARIMA is as follows.

$$y_{ij}=\varepsilon_{ij}+\psi_1\varepsilon_{i-1,j}+\psi_2\varepsilon_{i-2,j}+\ldots=\psi(B)\varepsilon_{ij} \tag{7}$$

Where $\psi(B)$ is determined by equation (8),

$$\theta(B)\nabla^d\psi(B)=\phi(B) \tag{8}$$

If the predictive step is 1, the predicted value of $\hat{y}_{ij}$ is,

$$\hat{y}_{ij}=\left(\psi_l\varepsilon_{ij}+\psi_{l+1}\varepsilon_{(i-1)j}+\psi_{l+2}\varepsilon_{(i-2)j}\cdots\right)+\left(\varepsilon_{(i+1)j}+\psi_1\varepsilon_{(i+l-1)j}+\ldots+\psi_{l-1}\varepsilon_{(i+1)j}\right) \tag{9}$$

(4) Residual and normal distribution test

The residual of performance parameters can be obtained by calculating the difference between the predicted values of performance parameters obtained from the ARIMA model and the measured value of the parameters. Normal distribution test can be conducted for the residual. If the residual complies with the normal distribution, the model is proved to be reasonable.

4.3. Reliability evaluation based on HMM

From the perspective of signal probability, HMM is categorized into Discrete Hidden Markov Model (DHMM) and Continuous Hidden Markov Model (CHMM) [8]. From the perspective of state transition the HMM is categorized into left-right HMM and traversal HMM. Because the performance degradation is an irreversible process, the left-right DHMM is adopted in this paper to describe the process of component performance degradation.

HMM is described as $\lambda = (N, M, \pi, A, B)$, where N is the number of hidden states, and the finite set of hidden states is defined as $S=\{S_1,S_2,\cdots,S_N\}$; M is the number of observations, and the finite set of observations is defined as $V=\{v_1,v_2,\cdots,v_M\}$; the probabilistic distributions of the initial states is $\pi=\{\pi_i\}$; A is the state transition probability matrix and $a_{ij}=P(q_t=j\,|\,q_{t-1}=i), 1\le i,j\le N$; B is the observation probability matrix and $B=(b_j(k)), b_j(k)=b_j(o_t)=(o_t=v_k\,|\,q_t=j)$.

4.3.1 Method for engineering application.

The real-time reliability evaluation of component of on-board equipment involves serveral performance degradation parameters. As the traditional HMM can only handle single observation time series for evaluating single parameter, it should be modified to be suitable for handling multiple observation series,. The process of the modified HMM multi-parameter estimation is as follows.

In order to use the multiple observation series to train the HMM, the equation of Baum-welch algorithm shall be revised. The set of observation series is defined as $O=\{O^{(1)},O^{(2)},\cdots,O^{(K)}\}$, and a observation series is defined as $O^{(k)}=\left\{O_1^{(k)},O_2^{(k)},\cdots,O_T^k\right\}$, and all the observation series are independent from each other. The probability of multiple observation series is $P(O\mid\lambda)=\sum_{k=1}^{K}\omega_k P\left(O^{(k)}\mid\lambda\right)$.The model parameter re-estimation equation and combinational weight of observation series are as follows:

$$\hat{\pi}_i=\sum_{k=1}^{K}\alpha_1^{(k)}(i)\beta_1^{(k)}(i)\Big/P\left(O^{(k)}\mid\lambda\right) \tag{10}$$

$$\hat{a}_{ij}=\frac{\sum_{k=1}^{K}\left[1/P_k\sum_{t=1}^{T_i-1}\alpha_t^{(k)}a_{ij}b_j\left(o_{t+1}^k\right)\beta_{t+1}^{(k)}(i)\right]}{\sum_{k=1}^{K}\left[1/P_k\sum_{t=1}^{T_i-1}\alpha_t^{(k)}(i)\beta_t^{(k)}(i)\right]} \tag{11}$$

$$\hat{b}_{jm}=\frac{\sum_{k=1}^{K}\left[1/P_k\sum_{t=1,o_t=v_m}^{T_i}\alpha_t^{(k)}(j)\beta_t^{(k)}(j)\right]}{\sum_{k=1}^{K}\left[1/P_k\sum_{t=1}^{T_i}\alpha_t^{(k)}(j)\beta_t^{(k)}(j)\right]} \tag{12}$$

$$\begin{cases}\omega_1=\frac{1}{K}p\left(O^{(2)}\mid O^{(1)},\lambda\right)\cdots\\ \qquad P\left(O^{(K)}\mid O^{(K-1)},\cdots,O^{(1)},\lambda\right)\\ \vdots\\ \omega_K=\frac{1}{K}p\left(O^{(1)}\mid O^{(K)},\lambda\right)\cdots\\ \qquad P\left(O^{(K-1)}\mid O^{(K)},O^{(K-2)}\cdots,O^{(1)},\lambda\right)\end{cases} \tag{13}$$

4.3.2 Real-time reliability calculation

The real-time reliability of the component is defined as the probability that the performance parameters are within the threshold range, namely the probability that the component is not in failure state. In order to calculate the probability of component in different states, conditional state probability $C(t)=\{C_1(t),C_2(t),\cdots,C_N(t)\}$ is introduced to describe the probability of component degradation state.

Given that the model parameter $\lambda^*=\left(A^*,B^*,\pi^*\right)$ and the observation series of *t-th* time $O_t=\{o_1,o_2,\cdots,o_t\}$ are known, the conditional state probability of time t in state i is defined as $C_i(t)=P\left(q_t=i\mid o_1,o_2,\cdots,o_N,\lambda^*\right)$, $q_t\in\{S_1,S_2,\cdots,S_N\}$, $t\geq 0, 1\leq i\leq N$. When $t=0$, $C(0)=\pi$.

The proof of the equation of conditional state probability is as follows.

Assume that the observation series of time t is denoted as O_t and the observation series of time t-1 is denoted as O_{t-1}. By definition, it can be known that:

$$C_j(t)=P\left(q_t=j\mid O_t,\lambda\right) \tag{14}$$

$$C_j(t)=\sum_{i=1}^{N}P\left(q_{t-1}=i,q_t=j\mid O_{t-1},o_t,\lambda\right) \tag{15}$$

$$C_j(t)=\sum_{i=1}^{N}\begin{bmatrix}\frac{P(q_{t-1}=i\mid O_{t-1},o_t,\lambda)}{P(o_t\mid O_{t-1},q_{t-1}=i,\lambda)}\times\cdots\\ \cdots P(q_t=j,o_t\mid O_{t-1},q_{t-1}=i,\lambda)\end{bmatrix} \tag{16}$$

$$\begin{aligned}&P(q_t=j,o_t\mid O_{t-1},q_{t-1}=i,\lambda)\\ &=P(q_t=j\mid O_{t-1},q_{t-1}=i,\lambda)\times\\ &\quad P(o_t\mid O_{t-1},q_{t-1}=i,\lambda,q_t=j)\\ &=b_j(o_t)\cdot a_{ij}=b_j(k)\cdot a_{ij}\end{aligned} \tag{17}$$

$$\begin{aligned}&\frac{P(q_{t-1}=i\mid O_{t-1},o_t,\lambda)}{P(o_t\mid O_{t-1},q_{t-1}=i,\lambda)}\\ &=\frac{P(q_{t-1}=i,o_t\mid O_{t-1},\lambda)}{P(o_t\mid O_{t-1},\lambda)}\cdot\frac{P(q_{t-1}=i\mid O_{t-1},\lambda)}{P(q_{t-1}=i,o_t\mid O_{t-1},\lambda)}\\ &=\frac{P(q_{t-1}=i\mid O_{t-1},\lambda)}{P(o_t\mid O_{t-1},\lambda)}=\frac{C_i(t-1)}{P(o_t\mid O_{t-1},\lambda)}\end{aligned} \tag{18}$$

According to equation (14) to (18), we can get

$$\begin{aligned}&P(o_t\mid O_{t-1},\lambda)=\sum_{i=1}^{N}\sum_{h=1}^{N}P(q_{t-1}=i,q_t=h,o_t\mid O_{t-1},\lambda)\\ &=\sum_{i=1}^{N}\sum_{h=1}^{N}\begin{bmatrix}P(q_t=h,o_t\mid O_{t-1},q_{t-1}=i,\lambda)\times\cdots\\ \cdots P(q_{t-1}=i\mid O_{t-1},\lambda)\end{bmatrix}\\ &=\sum_{i=1}^{N}\sum_{h=1}^{N}b_h(o_t)ai_hC_i(t-1)\end{aligned} \tag{19}$$

In conclusion,

$$\begin{aligned}C_j(t)&=\sum_{i=1}^{N}\frac{b_j(o_t)a_{ij}C_i(t-1)}{\sum_{i=1}^{N}\sum_{h=1}^{N}b_h(o_t)a_{ih}C_i(t-1)}\\ &=\frac{\sum_{i=1}^{N}C_i(t-1)a_{ij}b_j(o_t)}{\sum_{i=1}^{N}\sum_{h=1}^{N}C_i(t-1)a_{ih}b_h(o_t)}\end{aligned} \tag{20}$$

From equation (20), it can be seen that $\sum_{j=1}^{N}C_j(t)=1$. So the conditional state probability introduced in this paper can accurately describe the probability of different component states at any time.

From the C-K equation, it can be derived that $C'(t)=C(t)\cdot A$, where A is the state transition matrix at time t and $C(t)=(C_0(t),C_1(t),\cdots,C_k(t))$ is the conditional state vector. Laplace transform is used to calculate the real-time reliability, which is,

$$s\cdot\begin{bmatrix}C_0(s)\\ C_1(s)\\ \vdots\\ C_k(s)\end{bmatrix}-\begin{bmatrix}C_0(0)\\ C_1(0)\\ \vdots\\ C_k(0)\end{bmatrix}=\begin{bmatrix}a_{00}&0&\cdots&0\\ a_{01}&a_{11}&\cdots&0\\ \vdots&\vdots&\vdots&\vdots\\ a_{0k}&a_{1k}&\cdots&a_{kk}\end{bmatrix}\cdot\begin{bmatrix}C_0(s)\\ C_1(s)\\ \vdots\\ C_k(s)\end{bmatrix} \tag{21}$$

If the component in the initial state is under normal condition, there is the result of $C(0)$.

$$C(0)=\left(C_0(0),C_1(0),\cdots,C_k(0)\right)=(1,0,\cdots,0) \tag{22}$$

It can be seen from equation (21) that the probability of different component states at time t is derived by inverse Laplace transformation of $C(s)$. Therefor the reliability at time t is $R(t)=1-C_k(t)$.

5. BTM real-time reliability calculation

5.1. Selection of BTM performance parameters

BTM is one of the key components of the CTCS-3 on-board equipment, which has relatively high failure rate [9]. Therefore this paper takes BTM as an example to analyze the real-time reliability. The physical structure of BTM is shown as in figure 5, which consists of BTM host, transmission cable D and BTM antenna (CAU). A redundancy structure of double 2oo2 is adopted to improve the safety and reliability.

According to the BTM fault information collected by the data output system and the analysis of fault data, the performance parameters of BTM can be determined. Table 1 is the performance parameters corresponding to FMEA, which establishes the relationship between the FMEA and performance parameters. According to the requirements for choosing performance parameters described in section 3.1, the BTM self-test temperature, cable attenuation, train-track transmission power carrier frequency and bit error rate are chosen as the parameters for evaluation.

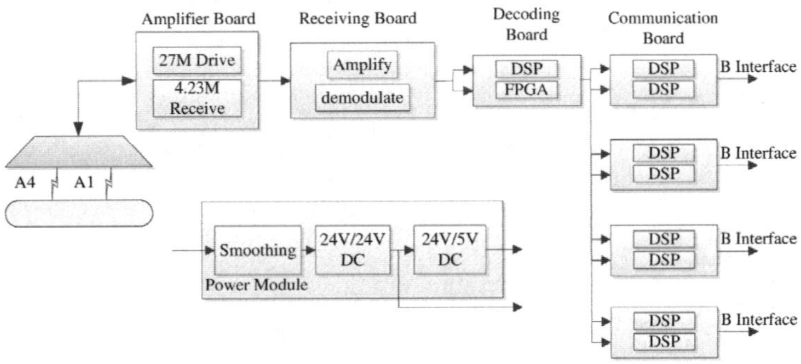

Figure 5. Structure of BTM.

5.2. ARIMA model prediction

The EVIEWS8.0 tool is used to build the ARIMA model. The sampling time is 20h. Sampling interpolation is adopted to improve the accuracy of fitting and prediction. A total of 120 data items are obtained, of which the first 100 data items are used for model training, and the other 20 data items are used for model validation. Figure 6 shows the track of the 4 performance parameters, where the empty circle stands for the measured value and the solid circle for the interpolated value. In the following BTM temperature is taken as an example to illustrate the ARIMA prediction process.

Table 1. Parameter Matching of BTM.

Functional Unit	Failure Mode	Parameter	Monitoring Point	Measuring Instrument
Power Board	Working indicator light flashing or go off	Voltage value	Power board pins 1、3	Digital Voltmeter
Amplifier Board, Receiver Board	BTM fails to complete self-test, the receiver indicator displays abnormally, Power-on self-test fault code 4-128-15-03	Temperature	BTM temperature self-test	Temperature Sensor
Cable D	BTM repeats self-test, fault code is 3-132-4、3-132-5	Vibration frequency	J3 line and P3 socket	Piezoelectric Acceleration Sensors
Communication	A communication board of	Cable attenuation	Cable	Attenuator

board	VC1 or DRU doesn't work			
BTM Antenna	Internal of cable contact is bad or short- circuited	Humidity; Electromagnetic interference of antenna port train-group transmission power	Antenna; antenna port	Humidity detector Electromagnetic interference tester Power meter
Decoding Board	Message Alarm	Bit Error rate	BTM decoding board	Bit Error rate analyser

Figure 7 is the Augmented-Dickey-Fuller(ADF) test for the original time series. It can be seen that the value of t is -1.95, which is much higher than the threshold value at the significance level of 1% - 10% level. So the original hypothesis is accepted, which means that the time series is non-stationary. After carrying out first order difference on the original time series, the result of ADF test for the obtained time series is shown in figure 8. It can be seen that the value of *t* is -3.999995, which is lower than the threshold value at the significance level of 1% . So the original hypothesis is rejected, which means the obtained time series is stationary. Figure 8 shows the result of ADF test for the obtained time series.

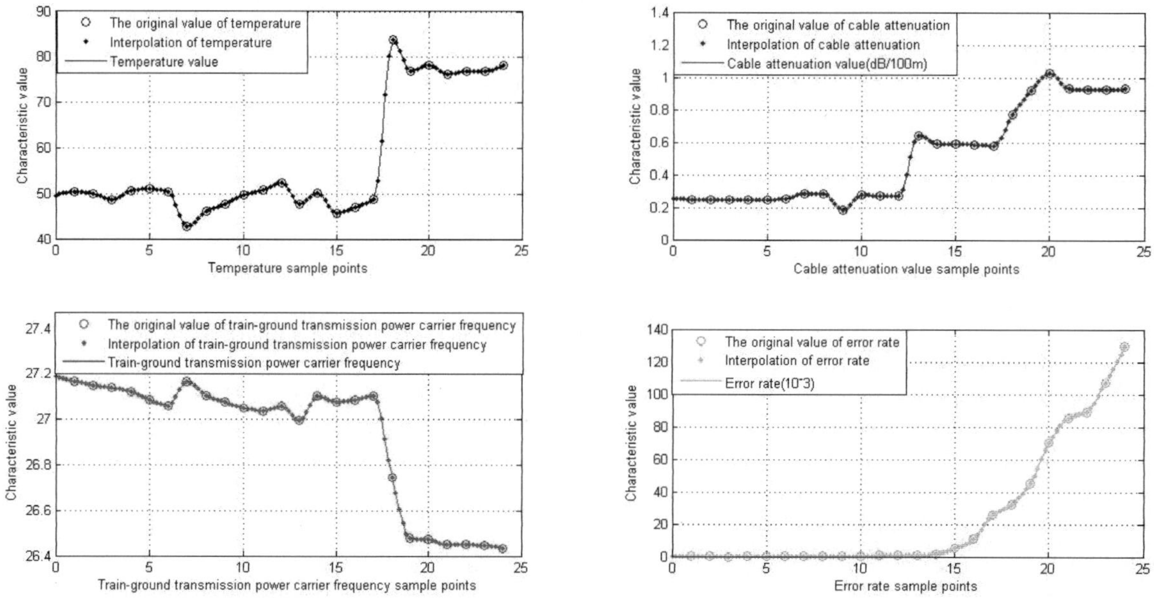

Figure 6. Measured value &interpolated value of performance parameters.

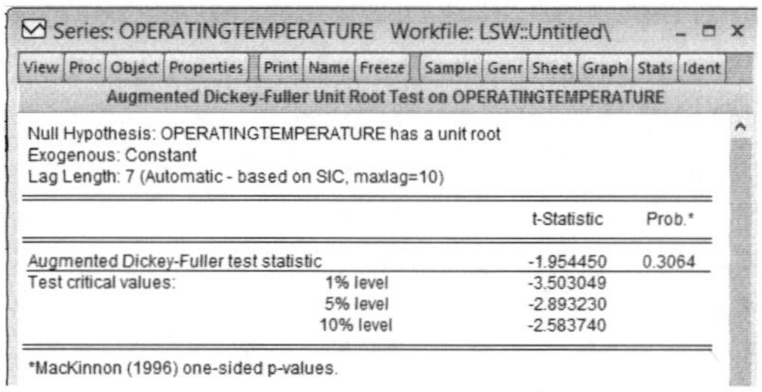
Series: OPERATINGTEMPERATURE Workfile: LSW::Untitled\

Augmented Dickey-Fuller Unit Root Test on OPERATINGTEMPERATURE

Null Hypothesis: OPERATINGTEMPERATURE has a unit root
Exogenous: Constant
Lag Length: 7 (Automatic - based on SIC, maxlag=10)

		t-Statistic	Prob.*
Augmented Dickey-Fuller test statistic		-1.954450	0.3064
Test critical values:	1% level	-3.503049	
	5% level	-2.893230	
	10% level	-2.583740	

*MacKinnon (1996) one-sided p-values.

Figure 7. ADF test for temperature seires.

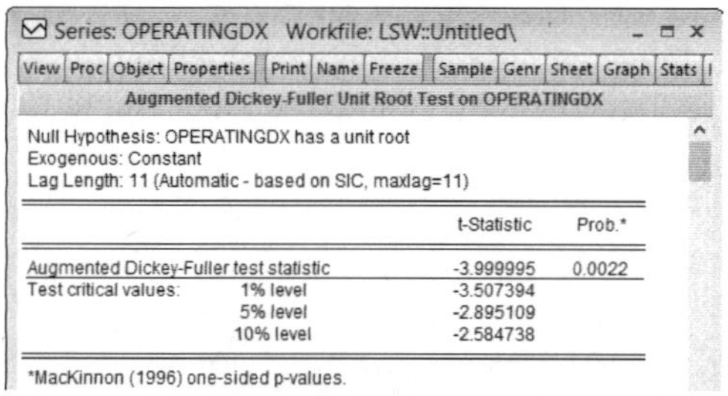
Series: OPERATINGDX Workfile: LSW::Untitled\

Augmented Dickey-Fuller Unit Root Test on OPERATINGDX

Null Hypothesis: OPERATINGDX has a unit root
Exogenous: Constant
Lag Length: 11 (Automatic - based on SIC, maxlag=11)

		t-Statistic	Prob.*
Augmented Dickey-Fuller test statistic		-3.999995	0.0022
Test critical values:	1% level	-3.507394	
	5% level	-2.895109	
	10% level	-2.584738	

*MacKinnon (1996) one-sided p-values.

Figure 8. ADF test for temperature series after first order difference.

Figure 9 shows the correlation analysis of the temperature series. It can be seen that all the values of Prob are less than 0.05, which suggests that the time series is correlated. By analyzing the autocorrelation and partial autocorrelation functions, it can be seen that the time series is suitable for the model of ARIMA (p, d, q)

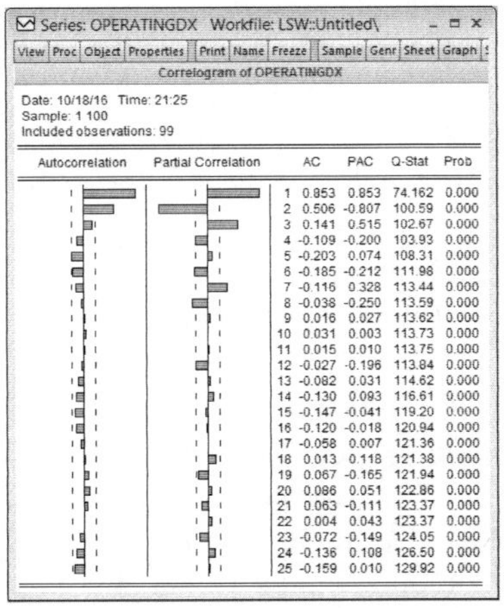
Series: OPERATINGDX Workfile: LSW::Untitled\
View Proc Object Properties Print Name Freeze Sample Genr Sheet Graph
Correlogram of OPERATINGDX

Date: 10/18/16 Time: 21:25
Sample: 1 100
Included observations: 99

	AC	PAC	Q-Stat	Prob
1	0.853	0.853	74.162	0.000
2	0.506	-0.807	100.59	0.000
3	0.141	0.515	102.67	0.000
4	-0.109	-0.200	103.93	0.000
5	-0.203	0.074	108.31	0.000
6	-0.185	-0.212	111.98	0.000
7	-0.116	0.328	113.44	0.000
8	-0.038	-0.250	113.59	0.000
9	0.016	0.027	113.62	0.000
10	0.031	0.003	113.73	0.000
11	0.015	0.010	113.75	0.000
12	-0.027	-0.196	113.84	0.000
13	-0.082	0.031	114.62	0.000
14	-0.130	0.093	116.61	0.000
15	-0.147	-0.041	119.20	0.000
16	-0.120	-0.018	120.94	0.000
17	-0.058	0.007	121.36	0.000
18	0.013	0.118	121.38	0.000
19	0.067	-0.165	121.94	0.000
20	0.086	0.051	122.86	0.000
21	0.063	-0.111	123.37	0.000
22	0.004	0.043	123.37	0.000
23	-0.072	-0.149	124.05	0.000
24	-0.136	0.108	126.50	0.000
25	-0.159	0.010	129.92	0.000

Figure 9. AC-PAC analysis of temperature series.

In the ACF and PACF analysis of the BTM temperature series after the first order difference, it can be seen that the 1-2 order autocorrelation coefficients are significant and the 3-25 order coefficients are not significant, so the value of q is initially set to 2; the 1-3 order partial autocorrelation coefficients are significant, so the value of p is initially set to 3. The values of AIC and SC are the most important criteria for choosing the model parameters p & q. In the optimal model, the values of AIC and SC are expected to be as small as possible. The model parameters of ARMA (3, 2) and ARMA (3, 3) are shown in figure 10 and figure 11 respectively. Table 2 is the value of AIC and SC for different p & q values.

From table 2 it can be seen that when the value of p and q are 3, the value of AIC and SC will both be the smallest. In figure 11 the model is ARMA (3, 3) and the R-squared is 0.959, which indicates

Table 2. ARMA Model Order Determination.

Model	AIC	SC
Eq03_02	1.11	1.27
Eq03_03	0.98	1.17
Eq03_01	1.30	1.44
Eq04_02	1.21	1.40
Eq04_03	1.22	1.43
Eq05_02	1.20	1.42
Eq05_03	1.22	1.47

Equation: UNTITLED Workfile: LSW::Untitled\
View Proc Object Print Name Freeze Estimate Forecast Stats Resids

Dependent Variable: OPERATINGDX
Method: Least Squares
Date: 10/18/16 Time: 21:33
Sample (adjusted): 5 100
Included observations: 96 after adjustments
Convergence achieved after 42 iterations
MA Backcast: OFF (Roots of MA process too large)

Variable	Coefficient	Std. Error	t-Statistic	Prob.
C	0.290010	0.151417	1.915305	0.0586
AR(1)	1.308137	0.735780	1.777891	0.0788
AR(2)	-0.536634	0.986439	-0.544012	0.5878
AR(3)	-0.088715	0.474775	-0.186857	0.8522
MA(1)	1.066470	0.712821	1.496125	0.1381
MA(2)	-0.071791	0.820591	-0.087487	0.9305

R-squared	0.952166	Mean dependent var	0.290858
Adjusted R-squared	0.949508	S.D. dependent var	1.822747
S.E. of regression	0.409577	Akaike info criterion	1.113079
Sum squared resid	15.09782	Schwarz criterion	1.273351
Log likelihood	-47.42781	Hannan-Quinn criter.	1.177864
F-statistic	358.3002	Durbin-Watson stat	1.781385
Prob(F-statistic)	0.000000		

Inverted AR Roots	.72-.45i	.72+.45i	-.12
Inverted MA Roots	.06	-1.13	

Estimated MA process is noninvertible

Figure 10. ARMA (3, 2) model.

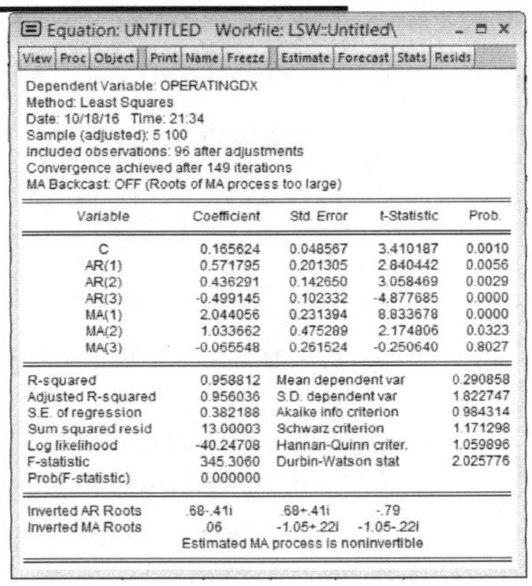
Equation: UNTITLED Workfile: LSW::Untitled\
View Proc Object Print Name Freeze Estimate Forecast Stats Resids

Dependent Variable: OPERATINGDX
Method: Least Squares
Date: 10/18/16 Time: 21:34
Sample (adjusted): 5 100
Included observations: 96 after adjustments
Convergence achieved after 149 iterations
MA Backcast: OFF (Roots of MA process too large)

Variable	Coefficient	Std. Error	t-Statistic	Prob.
C	0.165624	0.048567	3.410187	0.0010
AR(1)	0.571795	0.201305	2.840442	0.0056
AR(2)	0.436291	0.142650	3.058469	0.0029
AR(3)	-0.499145	0.102332	-4.877685	0.0000
MA(1)	2.044056	0.231394	8.833678	0.0000
MA(2)	1.033662	0.475289	2.174806	0.0323
MA(3)	-0.065548	0.261524	-0.250640	0.8027

R-squared	0.958812	Mean dependent var	0.290858
Adjusted R-squared	0.956036	S.D. dependent var	1.822747
S.E. of regression	0.382188	Akaike info criterion	0.984314
Sum squared resid	13.00003	Schwarz criterion	1.171298
Log likelihood	-40.24708	Hannan-Quinn criter.	1.059896
F-statistic	345.3060	Durbin-Watson stat	2.025776
Prob(F-statistic)	0.000000		

Inverted AR Roots	.68-.41i	.68+.41i	-.79
Inverted MA Roots	.06	-1.05+.22i	-1.05-.22i

Estimated MA process is noninvertible

Figure 11. ARMA (3, 3) model.

that the model is satisfactory. Except for the coefficient of MA (3), the other coefficients are significant. The temperature degradation model of BTM is determined as the ARMA (3, 3) model, which can be expressed as follows,

$$\begin{aligned}\hat{y}_t &= 0.1656 + 0.5718 y_{t-1} + 0.4363 y_{t-2} - 0.4991 y_{t-3} \\ &+ \varepsilon_t + 2.0441 \varepsilon_{t-1} + 1.0337 \varepsilon_{t-2} - 0.0655 \varepsilon_{t-3}\end{aligned} \tag{23}$$

Where $\hat{y}_t$ is the predicted value. White noise test is carried out for the model residuals, as shown in figure 12. It can be seen that the residuals of the autocorrelation and partial autocorrelation coefficients converge within 2 times the standard deviation. All the values of p are greater than 0.05, which suggests that the residuals are white noise process and the model is usable.

The fitting model ARIMA (3, 1, 3) is used for short-term prediction. Figure 13 is the prediction of ARIMA (3, 1, 3). It can be seen that the value of Theil is 0.1106. This value is small and it suggests that this model has an excellent prediction capability and the error ratio of the model is small. The variance proportion is 0.012339. It means that the prediction can simulate the fluctuation of the time series better. The covariance proportion is 0.987091. This value is big and it suggests that the result of prediction is desirable.

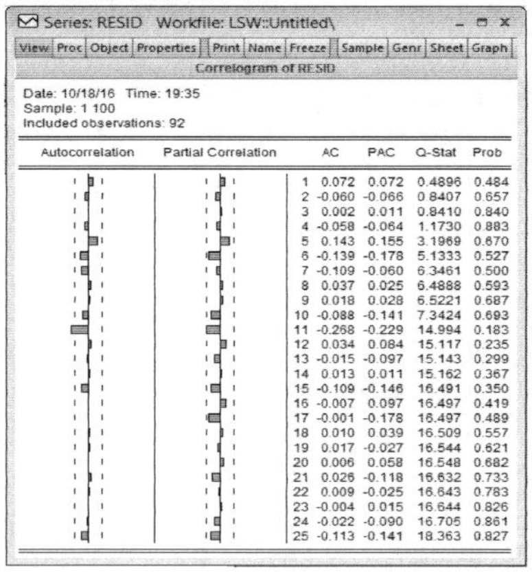

Figure 12. White noise test for residual.

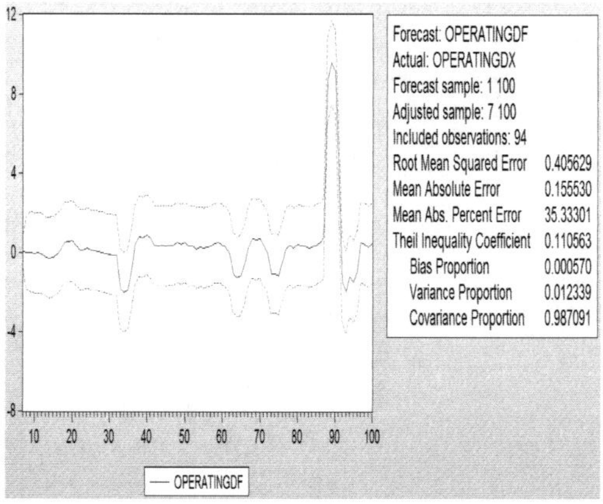

Figure 13. Static prediction of temperature series.

Similarly the performance degradation models of other parameters can be established.

(1)The performance degradation model of cable attenuation:

$$\hat{y}_t = 0.3241 y_{t-1} + 0.5639 y_{t-2} + \varepsilon_t - 0.9952 \varepsilon_{t-1} - 0.1513 \varepsilon_{t-2} \tag{24}$$

(2)The performance degradation model of train-ground transmission power carrier frequency:

$$\hat{y}_t = 0.3205 + 0.7312 y_{t-1} - 0.1163 y_{t-2} + 0.9622 y_{t-3} + \varepsilon_t - 0.2458 \varepsilon_{t-1} \tag{25}$$

(3)The performance degradation model of bit error rate:

$$\hat{y}_t = 1.1438 y_{t-1} - 0.1666 y_{t-2} + 0.4721 y_{t-3} + \varepsilon_t + 0.4735 \varepsilon_{t-1} - 0.1187 \varepsilon_{t-2} \tag{26}$$

5.3. Reliability calculation based on DHMM

The future data of the performance parameters can be obtained by the ARIMA model. DHMM is used to identify the probability of state transition. Firstly vector quantization is carried out for the obtained data from ARIMA, by using the performance status threshold matrix H and Lloyd algorithm. The calculation equation of the thresholds of performance parameters at each level is as follows:

$$S_i = S_0 + k_i \sigma \tag{27}$$

Where S_i is the state thresholds of the *i-th* level. S_0 is the factory value of performance parameters. σ is the standard deviation of performance parameters and k_i is the correction coefficient. Partition is determined by the performance status threshold matrix and the performance characteristic parameter series is divided into several sub-series by partition. These sub-series correspond with several consecutive integer values which are generally defined starting from 1. The index at different intervals is determined by the following equation:

$$Index\left(y_{ij}\right)=\begin{cases}1,\ y_{ij}\le partition\left(1,j\right)\\ i,\ partition\left(i-1,j\right)\le y_{ij}\le partition\left(i,j\right)\\ N,\ partition\left(n-1,j\right)\le y_{ij}\end{cases}\tag{28}$$

Where $partition\left(i,j\right)=h_{ij}$, h_{ij} is the value of performance status threshold matrix *H*. The matrix is defined as follows:

$$H=\left[h_{ij}\right]_{(n-1)\times m}=\begin{bmatrix}h_{11} & \cdots & h_{1m}\\ h_{21} & \cdots & h_{2m}\\ \cdots & \cdots & \cdots\\ h_{(n-1)1} & \cdots & h_{(n-1)m}\end{bmatrix}\tag{29}$$

The failure process of BTM can be regarded as a multi-state transition process. The state of BTM can be categorized into 4 types: normal state $\{0\}$, slight degradation state $\{1\}$, serious degradation state $\{2\}$ and failure state $\{3\}$ .According to experience and history data of parameters, the performance status threshold matrix of BTM is given by,

$$H=\begin{bmatrix}55 & 0.4 & 27.095 & 0.005\\ 60 & 0.6 & 25.095 & 0.05\\ 70 & 0.8 & 22.095 & 0.1\end{bmatrix}\tag{30}$$

The vector quantization set is the input observation series which need to be trained in DHMM. Baum Welch and EM algorithm is used to adjust the model parameters λ until the probability of observing the series $P(O_t\,|\,\lambda)$ is maximized, and then the required model is obtained. Set the initial model $\pi_0=\left[1,0,0,0\right]^T$. The state transition probability matrix A_0 and the observation probability matrix B_0 are set as follows.

$$A_0=\begin{bmatrix}0.8 & 0.1 & 0.06 & 0.04\\ 0 & 0.65 & 0.23 & 0.12\\ 0 & 0 & 0.63 & 0.37\\ 0 & 0 & 0 & 1\end{bmatrix}\tag{31}$$

$$B_0=\begin{bmatrix}0.25 & 0.25 & 0.25 & 0.25\\ 0.25 & 0.25 & 0.25 & 0.25\\ 0.25 & 0.25 & 0.25 & 0.25\\ 0.25 & 0.25 & 0.25 & 0.25\end{bmatrix}\tag{32}$$

The iterative training process is shown in figure 14. It can be seen that the log probability increases with the training times. After training for 10 times, the model tends to saturation. The parameters of HMM are as follows:

$$A=\begin{bmatrix}0.9641 & 0.0338 & 0.0021 & 0\\ 0 & 0.9266 & 0.0691 & 0.0043\\ 0 & 0 & 0.8976 & 0.1024\\ 0 & 0 & 0 & 1\end{bmatrix} \quad (33)$$

$$B=\begin{bmatrix}1 & 0 & 0 & 0\\ 1 & 0 & 0 & 0\\ 1 & 0 & 0 & 0\\ 0.1021 & 0.4290 & 0.1497 & 0.3193\end{bmatrix} \quad (34)$$

The real-time reliability computational equation can be derived from equation (21). The probability of BTM in different states is $C(t)=\left(C_0(t),C_1(t),C_2(t),C_3(t)\right)$.

$$C_0(t)=\exp(0.9641*t) \quad (35)$$

$$C_1(t)=0.901\exp(0.9641*t)-0.901\exp(0.9266*t) \quad (36)$$

$$C_2(t)=2.5321\exp(0.9641*t)-5.7340\exp(0.9266*t)+3.2019\exp(0.8976*t) \quad (37)$$

$$C_3(t)=1.1795\exp(0.8976*t)-2.1477\exp(0.9266*t)+0.9682\exp(0.9641*t) \quad (38)$$

The BTM real-time reliability is $R(t)=1-C_3(t)$, as shown in figure 15.

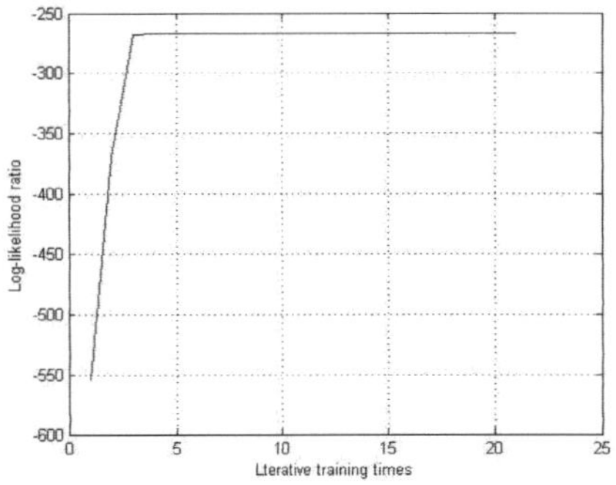

Figure 14. Iterative training process.

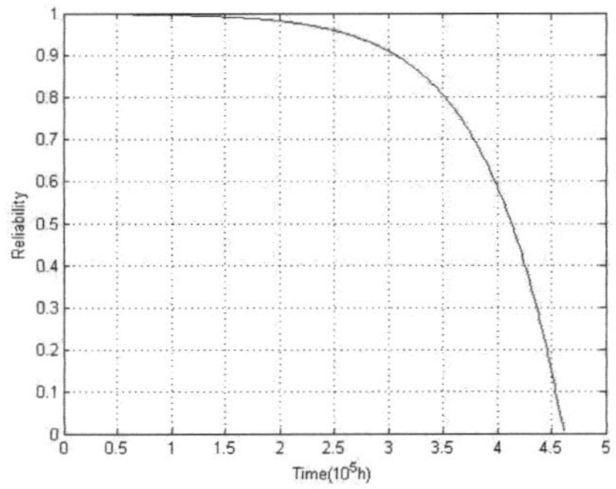

Figure 15. BTM real-time reliability.

6. Conclusion

On the basis of the traditional reliability analysis, this paper adopts the performance degradation parameters to estimate the real-time reliability of component of on-board equipment. ARIMA model and HMM are used to predict the trend of performance parameters and to calculate the real-time reliability. This method of real-time reliability evaluation can provide the support for conditioning based maintenance of on-board equipment of train control system.

7. References

[1] Transportation Bureau of the Ministry of Railways 2008 *The General Technical Scheme of CTCS-3 Train Control System* (Beijing: China Railway Press)

[2] Transportation Bureau of the Ministry of Railways 2009. *System Requirements Specification for CTCS-3 Train Control System (v1.0)*(Beijing: China Railway Press)

[3] Nagi G, Mark L, LI R and Jennifer R 2005 Residual-life distributions from component degradation signals: a Bayesian approach *Iie Trans.* **37** pp 543-57

[4] European Committee for Electrotechnical Standardization 2002 *EN 50129 Railway applications-Communication, signaling and processing systems safety related electronic*

systems for signaling.
[5] Shalev D M and Tiran J 2007 Condition-based fault tree analysis (CBFTA): a new method for improved fault tree analysis (FTA), reliability and safety calculations *Reliability Engineering & System Safety* **92** pp 1231-41
[6] Gawand H L, Murali N and Swaminathan P 2010 Reliability analysis by FMEA method for object oriented distributed digital control system design model of nuclear power plant *Conf. on Reliability, Safety and Hazard (ICRESH)* **37** pp 489-92
[7] Che J and Wang J 2010 Short-term electricity prices forecasting based on support vector regression and auto-regressive integrated moving average modeling *Energy Conversion & Management* **51** pp 1911-17
[8] Schuster-Böckler B and Bateman A 2007 An introduction to hidden Markov models *Current Protocols in Bioinformatics* pp 4-16
[9] Gao Z, Ma Z and Liang J 2014 Research on anti-electromagnet interference on BTM on-board equipment of CTCS3-300T train control system *Railway Signaling & Communication Engineering* **11** pp 1-6

Acknowledgement
This work was supported in part by the key project of Nature Science Foundation under Grant (61490705).

IOP Conf. Series: Materials Science and Engineering **351** (2018) 012005 doi:10.1088/1757-899X/351/1/012005

On circular m-consecutive-k,l-out-of-n:F systems

F S Makri

Department of Mathematics, University of Patras, Patras, Greece

makri@math.upatras.gr

Abstract. A general class of circular systems, called circular m-consecutive-k, l-out-of-n:F systems, is considered. For a system with independent and identically distributed component lifetimes, characteristics of the system's lifetime are studied via signature-based mixture representations. In deriving results combinatorial arguments are used.

1. Introduction

Since Kontoleon [1] introduced and studied the (linear) consecutive-k-out-of-n:F systems, many papers have appeared on consecutive-type systems. This is due to the fact that such systems have been used to model communication networks, oil pipeline systems, quality control inspection procedures, radar detection systems, etc. A linear (circular) consecutive-k-out-of-n:F system consists of n components ordered linearly (circularly) which fails if and only if there are k consecutive failed components. The circular system was introduced by Derman et al. [2]. Griffith [3] introduced and studied the m-consecutive-k-out-of-n:F system which is a system consisting of n components ordered linearly and fails if and only if there are m non-overlapping runs of k consecutive failed components. Alevizos et al. [4] considered the circular m-consecutive-k-out-of-n:F model for a system with circularly ordered components. Agarwal and Mohan [5] proposed and studied the m-consecutive-k-out-of-n:F system with overlapping runs. This system consists of n linearly ordered components and fails if and only if there are at least m overlapping runs of k consecutive failures. Eryilmaz and Mahmoud [6] introduced and studied the linear m-consecutive-k, l-out-of-n:F (m-lin/con/$(k, l)/n$:F) system that fails if and only if there are at least m l-overlapping runs of k consecutive failed components. The number of l-overlapping failure runs of length k is the number of failure runs of length k each of which may have overlapping (common) part of length at most l with the previous failure run of length k that has been enumerated.

The consecutive systems that have been studied are assumed to have components which function independently with the same probability, independently with different probabilities, dependently in a Markovian fashion or they are exchangeable dependent. Surveys on the subject can be found in Chao et al. [7], Kuo and Zuo [8], Eryilmaz [9] and Levitin [10].

In the paper, we study the circular m-consecutive-k, l-out-of-n:F system (m-cir/con/(k, l)/n:F), i.e., a system consisting of n circularly ordered components that fails if and only if there are at least m, $m \geq 1$, l-overlapping runs of k, $0 \leq l < k \leq n$, consecutive failed components. This system, as the corresponding linear system, via its two flexible parameters m and l, is on the one hand a generalization of all the above mentioned circular consecutive systems and on the other hand models new types of circular systems. For $m = 1$ it reduces to a circular consecutive-k-out-of-n:F system and for $l = 0$ it corresponds to a circular m-consecutive-k-out-of-n:F system.

m-cir/con/$(k,l)/n$:F was introduced by Makri and Psillakis [11] and studied using a run counting statistic $X_{n;k,l}^{(C)}$, $0 \le l < k \le n$, denoting the number of l-overlapping runs of failures of length k in a sequence $\{X_i\}_{i=1}^{n}$ of n two-state trials (failure-success) ordered on a circle. These authors also provided a potential application concerning an alarm or supervision system for a circular accelerator or for the core of a nuclear plant based on a m-cir/con/$(k,l)/n$:F system. The application takes into account the model's flexibility due to parameters m, l. The approach, instead, followed in the paper is based on the computation of mixture representations of the system lifetime distribution relied on certain sets of system's signatures. These mixtures have been proved to be useful tools in evaluating lifetime, and hence reliability, characteristics of a system and in comparing different systems, too.
In the signature-based mixture representations of a system consisting of n independent and identically distributed (i.i.d.), or more generally of exchangeable, component lifetimes crucial role has the vector $\boldsymbol{r}(n) = (r_1(n), r_2(n), \dots, r_n(n))$. In a n-component system, $r_i(n)$, $i = 1,2,\dots,n$, denotes the number of path sets of the structure with exactly i (working) components, and its computation can be relied on combinatorial arguments. Using these numbers, one can find explicitly signature-based representations (i.e. signatures, minimal and maximal signatures) for the system's structure and consequently, he/she can evaluate lifetime characteristics of the system. The theory on signature-based mixture representations of coherent systems has been developed for i.i.d. component lifetime distributions (see Samaniego [12], Navarro et al. [13] and the references therein) and is extended to exchangeable sequences by Navarro and Rychlik [14]. Further notable works on signatures are, among others, those of Boland [15], Eryilmaz [16], Eryilmaz et al. [17], Triantafyllou and Koutras [18], Koutras et al. [19] and Eryilmaz et al. [20].
In the paper, the lifetimes of system components are assumed to be i.i.d. In Section 2, we present some necessary preliminaries on signature-based mixture representations for the survival function (reliability) of a coherent system. These representations are then used to obtain results, presented in Section 3, on circular m-consecutive-k, l-out-of-n:F systems. Numerical examples illustrate further the theoretical results.

2. Signature-based mixture representations

The computation of system signatures is a counting problem and in essence, a combinatorial one (Samaniego [12]). To be more specific, let $T_1, T_2, \dots, T_n$, be the lifetimes of the components of a n-component coherent system that has lifetime T. Let us assume that $T_1, T_2, \dots, T_n$ are i.i.d. random variables from a continuous distribution, say F, on $(0,\infty)$. The signature of the system is defined as the vector $\boldsymbol{s} = \boldsymbol{s}(n) = (s_1, s_2, \dots, s_n)$ with

$$s_i = s_i(n) = P(T = T_{i:n}),\ \ i = 1,2,\dots,n, \tag{1}$$

where $T_{1:n} \le T_{2:n} \le \cdots \le T_{n:n}$ are the order statistics of the lifetimes $T_1, T_2, \dots, T_n$. $T_{i:n}$ represents the lifetime of an i-out-of-n failure system (i/n:F) that is, a system for which the i-th component failure causes its failure. Obviously, the signature $\boldsymbol{s}$ of a system is a probability vector, i.e. $s_i \ge 0, \sum_{i=1}^{n} s_i = 1$, since $P(T \in \{T_{1:n}, T_{2:n}, \dots, T_{n:n}\}) = 1$. As a counting problem the evaluation of the signature of a coherent system can be defined via (Kochar et al. [21])

$$s_i = n_i/n!\,, i = 1,2,\dots,n, \tag{2}$$

where n_i is the number of orderings of the component lifetimes for which the ith component failure causes system failure. That is, s_i represents the ratio of n_i to the (total) number of possible orderings, $n!$, of the n failure times $T_1, T_2, \dots, T_n$. Alternatively, it holds (Boland [15])

$$s_i(n) = \alpha_{n-i+1}(n) - \alpha_{n-i}(n) = \frac{r_{n-i+1}(n)}{\binom{n}{n-i+1}} - \frac{r_{n-i}(n)}{\binom{n}{n-i}},\ 1 \le i \le n, \tag{3}$$

where $\alpha_i = \alpha_i(n) = r_i(n)/\binom{n}{i}$, $i = 1,2,\dots,n$ ($\alpha_0(n)$=0, convention). $r_i = r_i(n)$, is the number of path sets of a $n-$component system with exactly i (working) components, and consequently α_i represents the proportion of them among $\binom{n}{i}$ such possible sets. Formula (3) is a fundamental tool for the computation of the signature of a coherent system since combinatorial arguments concerning the structure of the system can be used to derive the number of path sets of the system with i components.

Specifically, r_i can be determined by calculating the (total) number of binary sequences satisfying certain conditions which depend on the structure of the under study system.
Given the system's signature, the distribution of the system lifetime T, for i.i.d. component lifetimes with a distribution $F(\mathrm{t}) = \mathrm{P}(T_i \leq t), i = 1,2,\ldots,n,$ can be provided (Samaniego [12]) by

$$\bar{F}_T(t) = P(T > t) = \sum_{i=1}^{n} s_i\, P(T_{i:n} > t) \ = \sum_{i=1}^{n} s_i \sum_{j=0}^{i-1} \binom{n}{j} (F(t))^j (1 - F(t))^{n-j}. \tag{4}$$

Accordingly, the mean lifetime of the system can be expressed via its signature as

$$E(T) = \sum_{i=1}^{n} s_i\, E(T_{i:n}). \tag{5}$$

System signature is a useful tool in comparing systems in terms of stochastic ordering (see, Kochar et al. [21], Samaniego [12]). In brief, let $\boldsymbol{s}^{(u)} = (s_1^{(u)}, s_2^{(u)}, \ldots, s_n^{(u)}), u = 1,2,$ be the signatures of two coherent systems with respective lifetimes $T^{(u)}, u = 1,2,$ both based on n i.i.d. components with a common distribution F. Then,

$$\boldsymbol{s}^{(1)} \leq_{st} \boldsymbol{s}^{(2)} \text{ iff } \sum_{j=i}^{n} s_j^{(1)} \leq \sum_{j=i}^{n} s_j^{(2)}, i = 1,2,\ldots,n \tag{6}$$

and

$$\text{if } \boldsymbol{s}^{(1)} \leq_{st} \boldsymbol{s}^{(2)} \text{ then } T^{(1)} \leq_{st} T^{(2)}, \tag{7}$$

that is, $T^{(1)}$ is stochastically smaller than $T^{(2)}$, since the assumption $\boldsymbol{s}^{(1)} \leq_{st} \boldsymbol{s}^{(2)}$, implies

$$\bar{F}_{T^{(1)}}(t) = P(T^{(1)} > t) \leq P(T^{(2)} > t) = \bar{F}_{T^{(2)}}(t), \tag{8}$$

for all t, i.e. $T^{(1)}$ is less likely than $T^{(2)}$ to take values beyond t.
Alternatively, the survival function $\bar{F}_T(t)$ or the reliability function (polynomial for i.i.d. components) of the system can be written (since $\alpha_j = \sum_{i=n-j+1}^{n} s_i, j = 1,2,\ldots,n$) as

$$R(p(t)) = P(T > t) = \sum_{j=1}^{n} \alpha_j \binom{n}{j} p(t)^j q(t)^{n-j}, \tag{9}$$

where $p(t) = \bar{F}(t) = 1 - F(t) = 1 - q(t)$.

Furthermore, Eq. (4) can also be written as (see Appendix)

$$\bar{F}_T(t) = 1 - \sum_{i=1}^{n} s_i I_{F(t)}(i, n - i + 1), \tag{10}$$

where $I_x(v,\mu)$ is the incomplete Beta function ratio, given by $I_x(v,\mu) = \frac{\Gamma(v+\mu)}{\Gamma(v)\Gamma(\mu)} B(v;\mu;x)$ and $B(v;\mu;x) = \int_0^x t^{v-1}(1-t)^{\mu-1}, 0 \leq x < 1$, is the incomplete Beta function.

For n_f and n_w denoting the minimum number of failed components that may cause system failure and the maximum number of failed components such that the system can still work, respectively, it is true that

$$R(p) = \sum_{i=n-n_w}^{n} r_i(n) p^i (1-p)^{n-i} = 1 - \sum_{i=n_f}^{n_w+1} s_i(n)\, I_{1-p}(i, n - i + 1), \tag{11}$$

where $p = \bar{F}(t_0)$, for a fixed time t_0, and

$$E(T) = \sum_{i=n_f}^{n_w+1} s_i(n) E(T_{i:n}). \tag{12}$$

The numbers n_f and n_w depend on the under study system.
Navarro et al. [13] proved that any coherent system could be written as a generalized mixture of series or parallel systems. More specifically,

$$\bar{F}_T(t) = P(T > t) = \sum_{i=1}^{n} a_i P(T_{1:i} > t) = \sum_{i=1}^{n} \beta_i P(T_{i:i} > t), \tag{13}$$

where $\boldsymbol{a} = (a_1, a_2, \ldots, a_n)$ and $\boldsymbol{\beta} = (\beta_1, \beta_2, \ldots, \beta_n)$ are vectors, satisfying the conditions $\sum_{i=1}^{n} a_i = 1$, $\sum_{i=1}^{n} \beta_i = 1$ and are called, respectively, minimal and maximal signature of the system. Eryilmaz [16] proved that the minimal and maximal signature of any coherent system with exchangeable (and consequently with i.i.d.) components can be given as

$$a_i = \sum_{j=n-i}^{n_w} (-1)^{j-(n-i)} \binom{j}{n-i} r_{n-j}(n), \tag{14}$$

$$\beta_i = -\sum_{j=0}^{\min(i,n_w)} (-1)^{i-j} \binom{n-j}{n-i} r_{n-j}(n). \tag{15}$$

In (13), $T_{1:i}$ and $T_{i:i}$ are the lifetimes of an i-component series and a parallel system, respectively. Since,

$$P(T_{1:i} > t) = P(T_1 > t, T_2 > t, \dots, T_i > t) = (\bar{F}(t))^i \tag{16}$$

and

$$P(T_{i:i} > t) = 1 - P(T_1 \le t, T_2 \le t, \dots, T_i \le t) = 1 - (1 - \bar{F}(t))^i, \tag{17}$$

knowing the common component survival function $\bar{F}(t)$ and the number of path sets $r_i(n)$ the only that remains to determine the system's survival function $\bar{F}_T(t)$ is to specify n_w for the under study structure.

3. Circular m-consecutive-k, l-out-of-n:F systems

Lemma 1 (Makri et al. [22]). The number of allocations of γ indistinguishable balls into r distinguishable cells, i specified of which have capacity $m-1$ and each of the rest $r-i$ has capacity $n-1$ is given by

$$C_{i,r-i}(\gamma; m-1, n-1) = \sum_{j_1=0}^{\lfloor \frac{\gamma}{m} \rfloor} \binom{i}{j_1} \sum_{j_2=0}^{\lfloor \frac{\gamma - m j_1}{n} \rfloor} (-1)^{j_1+j_2} \binom{r-i}{j_2} \binom{\gamma - m j_1 - n j_2 + r - 1}{r-1}, \tag{18}$$

where $\lfloor x \rfloor$ denotes the greatest integer less than or equal to x.

Corollary 1 (Makri et al. [22]). Let $C_{i,r-i}(\gamma; m-1, n-1)$ be as in Lemma 1. Then,

$$C_{i,r-i}(\gamma; m-1, m-1) \equiv C(\gamma, r; m-1) = \sum_{j=0}^{\lfloor \frac{\gamma}{m} \rfloor} (-1)^j \binom{r}{j} \binom{\gamma - mj + r - 1}{r-1}. \tag{19}$$

The proofs of Theorem 1 and Proposition 1 are given in Appendix.

Theorem 1. The number, $r_i(n)$, of path sets with i elements of a circular m-consecutive-k, l-out-of-n:F system is given by the formula

$$r_i(n) = \sum_{x=0}^{m-1} r_{i,x}(n) \tag{20}$$

(convention: $r_0(n) = 0$) where

$$r_{i,0}(n) = \frac{n}{i} C(n-i, i; k-1) \tag{21}$$

and for $x \ge 1$,

$$r_{i,x}(n) = \frac{n}{i} \sum_{j=1}^{i} \binom{i}{j} \binom{x-1}{j-1} C_{j,i-j}(n-i-jk-(x-j)(k-l); k-l-1, k-1). \tag{22}$$

Proposition 1. For a circular m-consecutive-k, l-out-of-n:F system, it is true that for $m(k-l) \le n \le m(k-l)+l$, $n_f = n$, $n_w = n-1$ and for $n \ge m(k-l)+l+1$,

$$n_f = m(k-l) + l, \tag{23}$$

$$n_w = n - 2 - \left\lfloor \frac{n-1-n_f}{k} \right\rfloor. \tag{24}$$

Example 1. In Table 1, we depict the reliability $R_{k,m;n}^{(l)}(p)$, for $p = 0.50$ to 0.95, with step 0.05, of circular m-consecutive-k, l-out-of-n:F systems for $l = 0,1,2$ and $m = 2$, $k = 3$, $n = 10$. $R_{k,m;n}^{(l)}(p)$ can be computed using Eq. (11) or (13). The $r_i^{(l)}$, $s_i^{(l)}$, $a_i^{(l)}$ and $\beta_i^{(l)}$ are presented in Table 2. From Table 1, we observe that $R_{k,m;n}^{(l)}(p)$ does not increase with increasing l and is bounded from above by that of $l = 0$ and below by that of $l = 2 = k-1$. This feature is a common one for a m-cir/con/(k,l)/n:F with fixed m, n, k, p and varying l, $l = 0,1,\dots,k-1$.

Table 1. RELIABILITY $R_{3,2;10}^{(l)}(p), l = 0,1,2,$ of 2 −cir/con/(3, l)/10:F for $p = 0.50(0.05)0.95$

p/l	0	1	2
0.50	0.8623	0.7939	0.6670
0.55	0.9163	0.8656	0.7591
0.60	0.9532	0.9187	0.8360
0.65	0.9764	0.9551	0.8962
0.70	0.9895	0.9779	0.9401
0.75	0.9961	0.9907	0.9693
0.80	0.9989	0.9968	0.9867
0.85	0.9998	0.9992	0.9956
0.90	1.0000	0.9999	0.9991
0.95	1.0000	1.0000	0.9999

Table 2. PATH NUMBERS, $r_i^{(l)}$, SIGNATURES, $s_i^{(l)}$, MINIMAL, $a_i^{(l)}$ AND MAXIMAL, $\beta_i^{(l)}$ SIGNATURES, OF 2-CIR/CON/(3, l)/10: F SYSTEMS FOR $l = 0,1,2$

i	$r_i^{(l)}$	$a_i^{(l)}$	$\beta_i^{(l)}$	$s_i^{(l)}$	$r_i^{(l)}$	$a_i^{(l)}$	$\beta_i^{(l)}$	$s_i^{(l)}$	$r_i^{(l)}$	$a_i^{(l)}$	$\beta_i^{(l)}$	$s_i^{(l)}$
	$l = 0$				$l = 1$				$l = 2$			
1	0	0	0	0.0000	0	0	0	0.0000	0	0	0	0.0000
2	0	0	0	0.0000	0	0	0	0.0000	0	0	0	0.0000
3	59	60	0	0.0000	29	30	0	0.0000	9	10	0	0.0000
4	185	-235	0	0.0000	155	-55	0	0.0000	95	25	10	0.0476
5	251	402	0	0.0000	241	-58	10	0.0397	201	-158	-10	0.1508
6	210	-375	25	0.1190	210	275	5	0.2222	200	265	15	0.3492
7	120	200	-40	0.3810	120	-350	-30	0.4881	120	-210	-50	0.3691
8	45	-60	15	0.5000	45	220	5	0.2500	45	80	55	0.0833
9	10	10	0	0.0000	10	-70	20	0.0000	10	-10	-20	0.0000
10	1	-1	1	0.0000	1	9	-9	0.0000	1	-1	1	0.0000

In Table 2 we present, respectively, the vectors of the path numbers $\boldsymbol{r}^{(l)} = \left(r_1^{(l)}, r_2^{(l)}, \dots, r_n^{(l)}\right)$, the signatures $\boldsymbol{s}^{(l)} = \left(s_1^{(l)}, s_2^{(l)}, \dots, s_n^{(l)}\right)$, the minimal and the maximal $\boldsymbol{a}^{(l)} = \left(a_1^{(l)}, a_2^{(l)}, \dots, a_n^{(l)}\right)$ and $\boldsymbol{\beta}^{(l)} = \left(\beta_1^{(l)}, \beta_2^{(l)}, \dots, \beta_n^{(l)}\right)$, signatures, $0 \le l \le k-1$, of circular m-consecutive-k, l-out-of-n:F systems for $m = 2$, $k = 3$, $n = 10$. By the $s_i^{(l)}$ columns of Table 2 we easily verify, using Eq. (6), that $\boldsymbol{s}^{(2)} \le_{st} \boldsymbol{s}^{(1)} \le_{st} \boldsymbol{s}^{(0)}$. Therefore, for $T_{k,m;n}^{(l)}$ denoting the lifetime of a m-cir/con/(k, l)/n:F system, it holds $T_{3,2;10}^{(2)} \le_{st} T_{3,2;10}^{(1)} \le_{st} T_{3,2;10}^{(0)}$, which is consistent with the findings of Table 1, i.e. $R_{3,2;10}^{(2)}(p) \le R_{3,2;10}^{(1)}(p) \le R_{3,2;10}^{(0)}(p)$.

***Example** 2*. Let us consider that the component lifetimes $T_1, T_2, \dots, T_n$ are independent random variables with a common exponential distribution $F(t) = 1 - e^{-\lambda t}, t \ge 0$, and mean $1/\lambda, \lambda > 0$.

Then, the expected value of $T_{i:n}$ (i.e. of the i-th smallest component lifetime) is given by

$$E(T_{i:n}) = \lambda^{-1} \sum_{j=1}^{i} (n - j + 1)^{-1}, i = 1,2, \dots, n. \quad (25)$$

In Table 3, we compute, via Eq. (13), the survival function $\bar{F}_{T_{3,2;10}}^{(l)}(t)$, i.e. the reliability $R_{3,2;10}^{(l)}(t)$, of 2-cir/con/ $(3, l)/10$: F, for $l = 0,1,2$ and several values of t, when $\lambda = 1$. The $a_i^{(l)}$ and $\beta_i^{(l)}$ are presented in Table 2.

In Table 4, we compute, by Eqs. (12) and (25), the mean lifetime, $E(T_{k,m;n}^{(l)})$, i.e. the mean time to failure (MTTF), of circular m-consecutive-k, l-out-of-n:F systems, for several values of n, m, k and

$0 \leq l \leq k-1$, when $\lambda = 1$. From the Table, we observe that $E(T_{k,m;n}^{(l)})$ is increasing in m and k and decreasing in n and l.

Table 3. SURVIVAL FUNCTION $\bar{F}_{T_{3,2;10}}^{(l)}(t), l = 0,1,2$ OF 2-CIR/CON/$(3, l)/10$:F SYSTEMS, FOR EXPONENTIAL MODEL WITH $\lambda = 1$

$t \backslash l$	0	1	2
0.25	0.9980	0.9948	0.9806
0.50	0.9569	0.9243	0.8448
0.75	0.8242	0.7463	0.6108
1.00	0.6256	0.5235	0.3821
1.50	0.2652	0.1904	0.1131
2.00	0.0860	0.0547	0.0275
2.50	0.0239	0.0140	0.0062
3.00	0.0061	0.0034	0.0013

Table 4. MTTF, $E(T_{k,m;n}^{(l)}), l = 0,1, \ldots, k-1$, OF m-CIR/CON/$(k, l)/n$:F SYSTEMS

n	m	k	l	MTTF	n	m	k	l	MTTF
10	2	3	0	1.2325	20	3	3	0	1.0525
			1	1.1056				1	0.9381
			2	0.9389				2	0.7825
10	3	3	0	1.9290	20	2	4	0	1.1062
			1	1.5679				1	1.0565
			2	1.1512				2	0.9883
10	2	4	0	1.7623				3	0.8861
			1	1.6234					
			2	1.4448					
			3	1.2067					

4. Appendix

Proof of Equation (10). Since,

$$\sum_{j=i}^{n} \binom{n}{j} (F(t))^j (1-F(t))^{n-j} = \frac{n!}{(i-1)!\,(n-i)!} \int_0^{F(t)} u^{i-1} (1-u)^{(n-i+1)-1} du$$

$$= I_{F(t)}(i, n-i+1)$$

and $P(T > t) = 1 - \sum_{i=1}^{n} s_i \sum_{j=i}^{n} \binom{n}{j} (F(t))^j (\bar{F}(t))^{n-j}$, the result follows.

Proof of Theorem 1. Let $r_{i,x}(n)$ denote the number of circularly ordered binary sequences of n trials with i successes (working components) and x l-overlapping failure (failed component) runs of length k. Then, it is clear that $r_i(n) = r_{i,0}(n) + r_{i,1}(n) + \cdots + r_{i,m-1}(n)$. We will consider the evaluation of $r_{i,x}(n)$ mainly as a problem of allocation of balls into cells. The i working components displayed on the circle form i cells with a cell defined between two working components. These cells are made distinguishable by labelling them. The number of allocations of $n-i$ failures (failed components) into i cells so that no cell receives more than $k-1$ failures is $C(n-i, i; k-1)$, by Corollary 1, so that

$r_{i,0}(n) = \frac{n}{i} C(n-i, i; k-1)$, since the sequences are circularly arranged and each of the $C(n-i, i; k-1)$ arrangements gives n arrangements by rotation and the set of $nC(n-i, i; k-1)$

arrangements is partitioned into sets of i like arrangements. For $x \geq 1$ we observe that the number of allocations of $n-i$ failures into i distinguishable cells (created by the i working components) with j of them receiving x l-overlapping failure runs of length k and no one of the remaining $i-j$ cells receiving more than $k-1$ failures is $\binom{x-1}{j-1} C_{j,i-j}(n-i-jk-(x-j)(k-l); k-l-1, k-1)$. But the j cells (the ones containing the l-overlapping failure runs of length k) can be chosen in $\binom{i}{j}$ ways, $j = 1,2,\dots,min\left\{i, \left\lfloor\frac{n-i}{k}\right\rfloor\right\}$. So, the total number of allocations of the $n-i$ failed components in the i cells yielding x l-overlapping failure runs of length k is $\sum_{j=1}^{i}\binom{i}{j}\binom{x-1}{j-1} C_{j,i-j}(n-i-jk-(x-j)(k-l); k-l-1, k-1)$. Since the created sequences are circular, using the same arguments as above, the result for $r_{i,x}(n)$, $x \geq 1$, follows.

Proof of Proposition 1. For $n \geq l+m(k-l)+1$, it is clear that $l+m(k-l)$ consecutive failures (Fs) may cause system's failure while $l+m(k-l)-1$ can not, so that $n_f = l+m(k-l)$. Working in a similar way with that of Eryilmaz and Mahmoud [6] we consider one of the positions in the circular sequence as the first one and going clockwise we put $x = l+m(k-l)-1$ failures followed and preceded by a success (S), i.e., we have a sequence of the form

$$\underset{n}{\underline{S}}\,\underset{1}{\underline{F}}\,\underset{2}{\underline{F}}\cdots\underset{x}{\underline{F}}\,\underset{x+1}{\underline{S}}\,\overbrace{\cdots\cdots}^{n-x-2},$$

where the $n-x-2$ trials may include a maximum of $n-x-2-\left\lfloor\frac{n-x-2}{k}\right\rfloor$ failures so that the system can still work. Thus, $n_w = n-2-\left\lfloor\frac{n-l-m(k-l)-1}{k}\right\rfloor$. For $m(k-l) \leq n \leq l+m(k-l)$ the results for n_f and n_w are straightforward.

5. References

[1] Kontoleon J M 1980 Reliability determination of a r-successive-out-of-n:F system *IEEE Trans. Reliab.* vol. 29, pp 437-438

[2] Derman C, Lieberman G J and Ross S M 1982 On the consecutive-k-out-of-n:F system *IEEE Trans. Reliab.* vol. 31, pp 57-63

[3] Griffith W S 1986 On consecutive k-out-of-n failure systems and their generalizations *Reliability and Quality Control,* (A.P. Basu Ed., Amsterdam: Elsevier) pp 157-165

[4] Alevizos P D, Papastavridis S G and Sypsas P 1993 Reliability of cyclic m-consecutive-k-out-of-n:F system *Proceedings of IASTED Conference on Reliability, Quality Control, and Risk Assessment,* (H. Pham and M.H. Hamza Eds., Calgary: ACTA Press) pp 140-143

[5] Agarwal M, Mohan P 2008 GERT analysis of m-consecutive-k-out-of-n:F system with overlapping runs and $(k-1)$-step Markov dependence *Int. J. Operat. Res.* vol. 3, pp 36-51

[6] Eryilmaz S, Mahmoud B 2012 Linear m-consecutive-k,l-out-of-n:F system *IEEE Trans. Reliab.* vol. 61, pp 787-791

[7] Chao M T, Fu J C and Koutras M V 1995 A survey of the reliability studies of consecutive-k-out-of-n:F system and its related systems *IEEE Trans. Reliab.* vol. 44, pp 120-127

[8] Kuo W, Zuo M 2003 *Optimal Reliability Modeling: Principles and Applications* (Hoboken, NJ:Wiley)

[9] Eryilmaz S 2010 Review of recent advances in reliability of consecutive-k-out-of-n and related systems *Proc. Instit. Mech. Eng. Part O: J. Risk Reliab.* vol. 224, pp 225-237

[10] Levitin G 2010 *The Universal Generating Function in Reliability Analysis and Optimization* (London: Springer-Verlag Limited)

[11] Makri F S, Psillakis Z M 2015 On l-overlapping runs of ones of length k in sequences of independent binary random variables *Commun. Statist. Theor. Meth.* vol. 44, pp 3865-3884

[12] Samaniego F J 2007 *System Signatures and their Applications in Engineering Reliability* (New York: Springer)

[13] Navarro J, Ruiz J M and Sandoval C J 2007 Properties of coherent systems with dependent components *Commun. Statist. Theor. Meth.*, vol. 36, pp 175-191

IOP Conf. Series: Materials Science and Engineering **351** (2018) 012005 doi:10.1088/1757-899X/351/1/012005

[14] Navarro J, Rychlik T 2007 Reliability and expectation bounds for coherent systems with exchangeable components *J. Multivariate Anal.* vol. 98, pp 102-113
[15] Boland P 2001 Signatures of indirect majority systems *J. Appl. Probab.* vol. 38, pp 597-603
[16] Eryilmaz S 2010 Mixture representations for the reliability of consecutive-k systems *Math. Comput. Model.* vol. 51, pp 405-412
[17] Eryilmaz S, Koutras M V and Triantafyllou I S 2011 Signature based analysis of m-consecutive-k-out-of-n:F systems with exchangeable components *Naval Res. Logist.* vol 58, pp 344-354
[18] Triantafyllou I S, Koutras M V 2014 Reliability properties of (n, f, k) systems *IEEE Trans. Reliab.* vol. 63, pp 357-366
[19] Koutras M V, Triantafyllou, I S and Eryilmaz S 2016 Stochastic comparisons between lifetimes of reliability systems with exchangeable components *Method. Comput. Appl. Probab.* vol. 18, pp 1081-1095
[20] Eryilmaz S, Koutras M V and Triantafyllou I S 2016 Mixed three-state k-out-of-n systems with components entering at different performance levels *IEEE Trans. Reliab.* vol. 65, pp 969-972
[21] Kochar S, Mukerjee H and Samaniego F J 1999 The 'signature' of a coherent system and its application to comparisons among systems *Naval Res. Logist.* vol. 46, pp 507-523
[22] Makri F S, Philippou A N and Psillakis Z M 2007 Polya, inverse Polya, and circular Polya distributions of order k for l-overlapping success runs *Commun. Statist. Theor. Meth.* vol 36, pp 657-668

IOP Conf. Series: Materials Science and Engineering **351** (2018) 012006 doi:10.1088/1757-899X/351/1/012006

Environmental sustainability assessment of hydropower plant in Europe using life cycle assessment

M A P Mahmud, N Huda, S H Farjana, and C Lang

School of Engineering, Macquarie University, Sydney, NSW 2109, Australia

nazmul.huda@mq.edu.au

Abstract. Hydropower is the oldest and most common type of renewable source of electricity available on this planet. The end of life process of hydropower plant have significant environmental impacts, which needs to be identified and minimized to ensure an environment friendly power generation. However, identifying the environmental impacts and health hazards are very little explored in the hydropower processing routes despite a significant quantity of production worldwide. This paper highlight the life-cycle environmental impact assessment of the reservoir based hydropower generation system located in alpine and non-alpine region of Europe, addressing their ecological effects by the ReCiPe and CML methods under several impact-assessment categories such as human health, ecosystems, global warming potential, acidification potential, etc. The Australasian life-cycle inventory database and SimaPro software are utilized to accumulate life-cycle inventory dataset and to evaluate the impacts. The results reveal that plants of alpine region offer superior environmental performance for couple of considered categories: global warming and photochemical oxidation, whilst in the other cases the outcomes are almost similar. Results obtained from this study will take part an important role in promoting sustainable generation of hydropower, and thus towards environment friendly energy production.

1. Introduction

In this modern era, world has experienced rising economic growth and increasing energy demands that make happen detrimental effects on the environment [1]. These environmental effects, the outcome of the improper practice of available assets, have a worldwide level in few circumstances; for instance global warming, climate change and so on. The major consequences of global warming are temperature increase and ocean's surface escalation. Human activities, such as deforestation, land use changes, agriculture and combustion of fossil fuels [2]. The combustion of coals or gases for power production and transportation induce release of carbon dioxide (CO_2), nitrous oxides and greenhouse gases (GHG) like chlorofluorocarbons (CFCs) and methane, which are the primary providers to the greenhouse effect. The development of environment friendly, sustainable hydropower generation system is gaining substantial attention due to its unrivalled potential for extensive renewable energy production and GHG release diminution.

Considering the noteworthy role that hydropower production technology has to take part in meeting today's power crisis, researchers require to assess the environmental life cycle impact analysis considering different stages of hydroelectricity production technology. The life cycle assessment (LCA) technology assists to assess the total GHG emissions causing from the consumption of hydropower as a source of energy. LCA is a systematic approach which gives an evaluation of the net environment-related effects of the products from a cradle-to-grave schemes employing the

comprehensive life cycle input elements and output emissions which work under the specific life cycle inventory (LCI) [3-6].
Though water itself is not lethal, the electricity production process involves many stages, which creates environmental issues. Furthermore, the transportation medium of these elements to the plant location releases hazardous particles i.e., carbon monoxide, dust and carcinogenic particles. As Europe has the largest resource of renewable energy generation through hydropower, it is quite expected that both alpine and non-alpine regions of Europe is facing the environmental impacts and health effects caused by the hydropower generation process.
All the hydropower generation based LCA research until now have addressed on the assessment of the GHG releases. No prior LCA-based researches [7-9] assessed and compared the total environmental impacts by hydropower production technologies in Europe. This work purposes to fill this research gap by executing a complete environmental LCA of hydropower production technologies in Europe. This research paper, will highlight the environmental threats associated with hydropower production by a systematic LCA method considering whole life period of the plant. We also compare the environmental impacts for two different cases of hydropower plants in Europe under various assessing categories. The environmental impacts of hydropower generation process in alpine and non-alpine regions of Europe through LCA by comparing their generation stages are addressed and the outcomes at different assessing categories are analyzed. Section 2 provides a comprehensive outline of the environmental life cycle impact assessment methods. Section 3 highlights the LCA outcomes under the ecological effect assessment categories, using the ReCiPe technique and the CML technique, thus recommends the environmental-effect reduction methods for selected processes.

2. Methodology

Hydropower generation take place harvesting mechanical energy from the flowing water using a turbine or rotor. The main elements which assist to produce energy by hydropower plants are the alternator, the hydraulic turbine, the transformers, and the reservoir [3]. A turbine constitutes of several blades joined to a moving shaft. The speedy contact between turbine blades and flowing water create a movement of the shaft. Then the alternator transforms the mechanical power from the moving shaft to electricity. The transformers work as a connecting route between the alternator and consumer through wires. The reservoirs hoard water prior to the flow via blades.
The aim of the study is to analyze the environmental impact of two different hydropower plant located in alpine and non-alpine regions of Europe, and compare their environmental impacts under several impact categories. The environmental impact calculation is the sub-section of a LCA analysis in which the outcome for the life cycle inventory (LCI), the general flows of each step, are converted into LCA category indicator outcomes [2]. This is done using characterization simulations. The characterization simulations included in the CML-IA (baseline) methodology [10, 2] were utilized in this research work. Six different life cycle impact categories were measured: abiotic depletion, global warming, ozone layer depletion, photochemical oxidation, acidification and eutrophication. Simapro 8.4 [11] software package was used to perform the life cycle inventory and impact assessment calculations [2]. The scope datasets of this research work are gathered from Australian Life Cycle Inventory (AusLCI) database.
The investigation approaches are selected as the ReCiPe technique and the CML technique, which are generally applied to electricity generation technology-based environmental impact evaluations. The approach of ecological effect assessment of hydropower generation system is cradle-to-grave. These datasets are renewable energy technology related where the major emissions are distributed among electricity generation parts. Table 1 shows the comparative input and output datasets of the reservoir based hydropower plant process at alpine and non-alpine regions. It highlights that the input rates are same for both alpine and non-alpine location based hydropower plants at all the categories .It also shows that the hazardous waste incineration associated with production activities such as mineral oil disposal and biogenic methane. Although mineral oil disposal is same for both systems but non-alpine region power plants shows higher biogenic methane emission compared to alpine region plants

Table 1. Life-cycle inputs and outputs of hydropower plant at alpine and non-alpine regions in Europe

Inputs and Outputs	Amount		Unit
	Electricity, hydropower, at reservoir power plant, alpine region/RER U/AusSD U	Electricity, hydro-power, at reservoir power plant, non alpine regions/RER U/AusSD U	
Inputs			
Transformation, from unknown	0.000023	0.000023	m^2
Transformation, to water bodies, artificial	0.0000228	0.0000228	m^2
Transformation, to industrial area, built up	0.00000023	0.00000023	m^2
Occupation, water bodies, artificial	0.0035	0.0035	m^2a
Volume occupied, reservoir	0.15	0.15	m^3y
Water, turbine use, unspecified natural origin	0.81	0.81	m^3
Energy, potential (in hydropower reservoir), converted	3.79	3.79	MJ
Lubricating oil, at plant/RER U/AusSD U	0.000007	0.000007	kg
Reservoir hydropower plant, alpine region/RER/I U/AusSD U	$3.90E^{-13}$	$3.90E^{-13}$	p
Outputs			
Methane, biogenic	0.000014	0.000286	kg
Disposal, used mineral oil, 10% water, to hazardous waste incineration/CH U/AusSD U	0.000007	0.000007	kg

3. Results and Discussion

The outcomes are represented at impact evaluation aspect. As it is a relative analysis, the finding for every impact type for the system with the maximum effects are counted 100 percent and the outcome for another system is counted as the corresponding ratio.

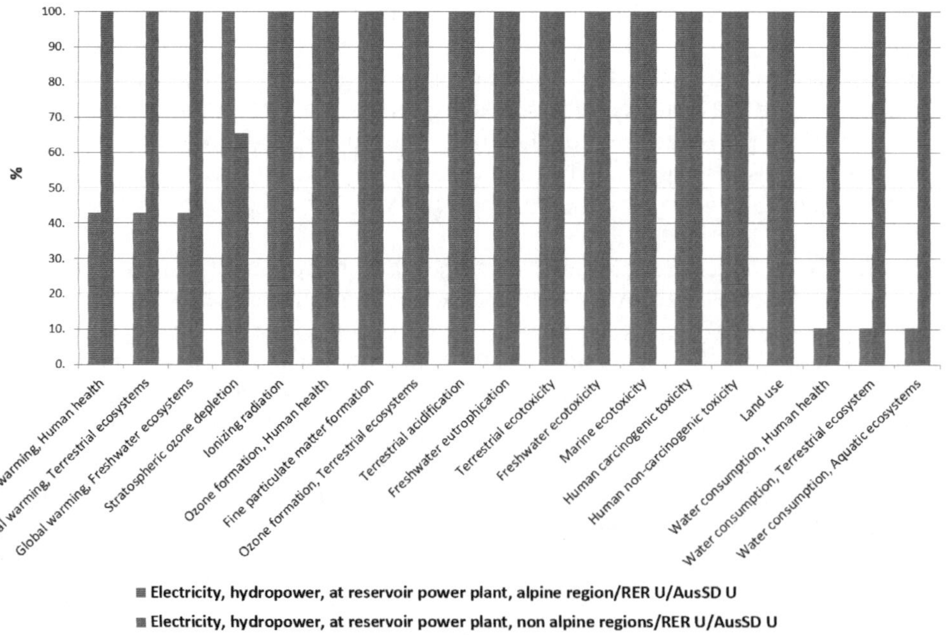

Figure 1. Impact comparison using ReCiPe 2016 Endpoint methodology.

Figures 1 to 3 illustrate the comparative environmental impacts caused by hydropower plant technologies at alpine and non-alpine regions in Europe. Figure 1 shows the environmental impact comparison applying ReCiPe 2016 Endpoint methodology. It is transparent that plants of alpine region perform better environmentally than non-alpine region except in one impact category, stratospheric ozone depletion. The endpoint comparison after weighing, using ReCiPe 2016 methodology is represented at Figure 2. As shown, hydropower plant in non-alpine regions results in more damage compared to alpine regions.

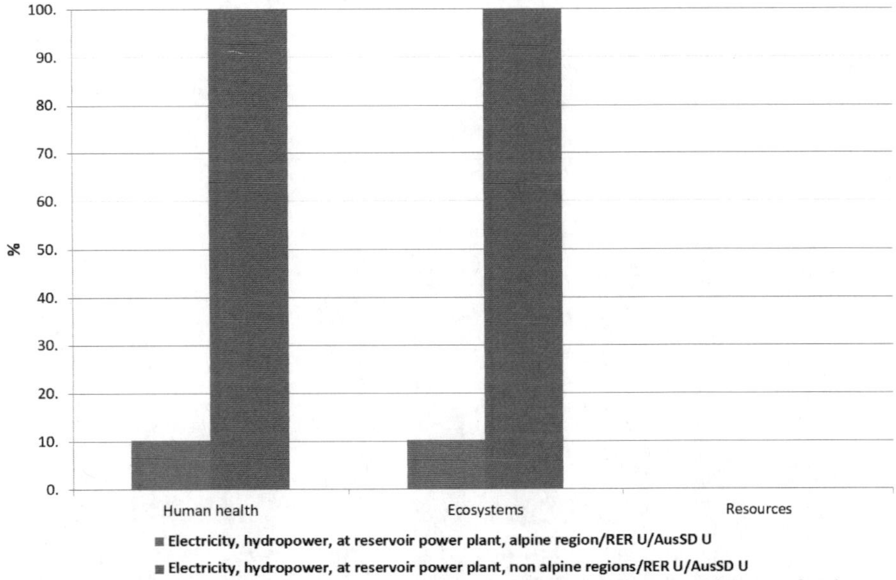

Figure 2. Endpoint comparison after weighing, using ReCiPe method.

Figure 3 illustrates the comparative LCA-based environmental effect evaluation of hydropower plant in alpine and non-alpine regions of Europe using CML method. The results indicate that plants of alpine region offer superior action for couple of studied impact groups: global warming and photochemical oxidation. Although in the other cases the outcomes are almost similar. This analysis highlight the LCA-based environmental influences caused by the hydropower plant systems in Europe, to get the queries arising from ecological threats.

Table 2. Life-cycle impact assessment outcomes-ReCiPe method.

Impact category	**Electricity, hydropower, at reservoir power plant, alpine region/RER U/AusSD U**	**Electricity, hydro-power, at reservoir power plant, non-alpine regions/RER U/AusSD U**	**Unit**
Global warming, Human health	1.7838999 E-9	4.1655307 E-9	DALY
Global warming, Terrestrial ecosystems	5.3818149 E-12	1.2556865 E-11	species.yr
Global warming, Freshwater ecosystems	1.4702874 E-16	3.4298551 E-16	species.yr
Stratospheric ozone depletion	3.6271486 E-13	2.3780375 E-13	DALY
Ionizing radiation	1.0769103 E-13	1.0769103 E-13	DALY
Ozone formation, Human health	4.8714449 E-12	4.8714448 E-12	DALY
Fine particulate matter formation	1.2586791 E-9	1.258679 E-9	DALY

Ozone formation, Terrestrial ecosystems	7.0163971 E-13	7.016397 E-13	species.yr
Terrestrial acidification	1.1623602 E-12	1.1623602 E-12	species.yr
Freshwater eutrophication	3.762101 E-14	3.762101 E-14	species.yr
Terrestrial ecotoxicity	5.5926631 E-14	5.592663 E-14	species.yr
Freshwater ecotoxicity	1.4610729 E-15	1.4610729 E-15	species.yr
Marine ecotoxicity	5.143076 E-16	5.143076 E-16	species.yr
Human carcinogenic toxicity	5.2830406 E-10	5.2830405 E-10	DALY
Human non-carcinogenic toxicity	2.499131 E-11	2.4991309 E-11	DALY
Land use	3.7055138 E-13	3.7055138 E-13	species.yr
Water consumption, Human health	5.0869534 E-7	5.0041953 E-6	DALY
Water consumption, Terrestrial ecosystem	3.0934176 E-9	3.0430917 E-8	species.yr
Water consumption, Aquatic ecosystems	1.3840179 E-13	1.3615018 E-12	species.yr

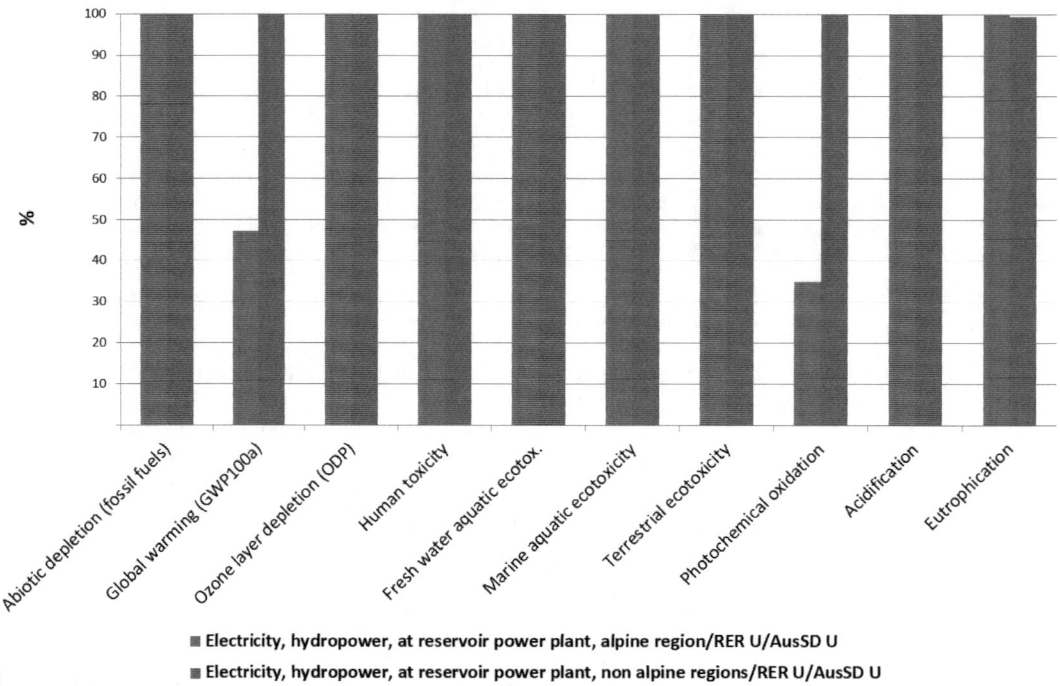

Figure 3. Comparative ecological influence evaluation of hydropower plant in alpine and non-alpine regions of Europe using CML method.

The assessment is performed utilising the ReCiPe technique and the CML technique considering various ecological influence groups which could be generally categorized as – global warming, ozone formation, acidification, eutrophication, ecotoxicity, human toxicity, water consumption, stratospheric ozone depletion, ionizing radiation, land use, etc [10]. Among the key impact groups which are exemplified at tables 2 and 3, the whole outcomes show that a substantial ecological influence is occurred by the non-alpine region plants over the alpine region plants. The reason behind this is that the long distance transportation of raw materials in non-alpine region hydropower plants due to unavailability at nearby locations where raw materials of the alpine based plants is available at nearby locations.

IOP Conf. Series: Materials Science and Engineering **351** (2018) 012006 doi:10.1088/1757-899X/351/1/012006

Table 3. Life-cycle impact assessment outcomes-CML method.

Impact category	**Electricity, hydropower, at reservoir power plant, alpine region/RER U/AusSD U (%)**	**Electricity, hydropower, at reservoir power plant, non alpine regions/RER U/AusSD U (%)**
Abiotic depletion (fossil fuels)	100	100
Global warming (GWP100a)	47.2216	100
Ozone layer depletion (ODP)	100	100
Human toxicity	100	100
Fresh water aquatic ecotox.	100	100
Marine aquatic ecotoxicity	100	100
Terrestrial ecotoxicity	100	100
Photochemical oxidation	35.0314	100
Acidification	100	100
Eutrophication	100	99.4055

Table 4 shows the contrast per impact type, by addition of each effects, using Ecoindicator 2016 methodology. It is transparent that plants of alpine region perform better environmentally than non-alpine region plants. The comparative ecological influence evaluation of hydropower plants in alpine and non-alpine regions of Europe using Ecoindicator approach is represented at Figure 4.

Table 4. Life-cycle impact assessment results by Ecoindicator approach

Impact category	**Electricity, hydro-power, at reservoir power plant, alpine region/RER U/AusSD U**	**Electricity, hydro-power, at reservoir power plant, non-alpine regions/RER U/AusSD U**	**Unit**
Human non-carcinogenic toxicity	2.499131 E-11	2.4991309 E-11	DALY
Human carcinogenic toxicity	5.2830406 E-10	5.2830405 E-10	DALY
Fine particulate matter formation	1.2586791 E-9	1.258679 E-9	DALY
Ionizing radiation	1.0769103 E-13	1.0769103 E-13	DALY
Stratospheric ozone depletion	3.6271486 E-13	2.3780375 E-13	DALY
Freshwater eutrophication	3.762101 E-14	3.762101 E-14	species.yr
Terrestrial acidification	1.1623602 E-12	1.1623602 E-12	species.yr
Land use	3.7055138 E-13	3.7055138 E-13	species.yr

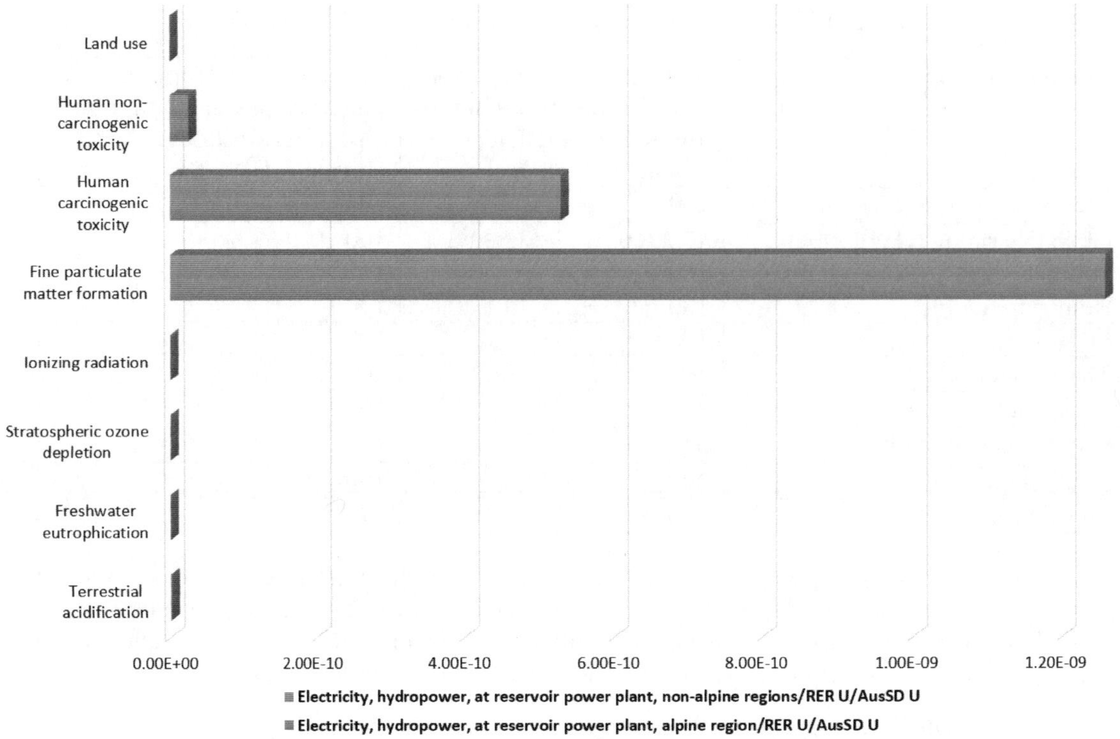

Figure 4. Comparative ecological influence evaluation of hydropower plant in alpine and non-alpine regions of Europe using Ecoindicator method.

The results reveal that the maximum impact is occurred at fine particulate matter formation impact category. Whilst the minimum is obtained freshwater eutrophication category by both types of hydropower plants. The reason behind these impacts is the amount of toxic materials present as constituent of plant structure and its electricity production steps.
On the whole, the outcomes obtained from this comparative analysis show that a substantial ecological danger is generated from the hydropower plants in Europe, which will be greater in the upcoming days due to reduction of quality raw materials. Thus will release higher toxic elements to the environment, which are required to be controlled to save the earth and its creatures.

4. Conclusion

Europe has the abundant source of a hydropower plant, the biggest in the entire world. By providing environmental sustainability, Europe could generate additional hydropower from now onwards, which could further reduce the ever-growing energy demand. To generate energy in an environmentally-safe procedure, attention should be given in making the raw material extraction systems optimized and sustainable. In this paper, we highlight the comparative environmental impact of hydropower plants located in alpine and non-alpine region of Europe using the CML and ReCiPe methods utilizing AusLCI database and SimaPro software. The results indicate that plants of alpine region have a better performance over non-alpine cases. This research outcome will play a vital role in promoting environment friendly and sustainable generation of hydropower.

5. References

[1] Tsoutsos T, Frantzeskaki N and Gekas V 2005 Environmental impacts from the solar energy technologies, Energy Policy 33, pp 289-296
[2] Ramirez A D, Boero A J and Melendres A M 2015 Using life cycle assessment to compare the environmental performance of electricity generation technologies in Ecuador: fuel oil in internal combustion engines vs. fuel oil in steam power plants, WIT Transaction on Ecology and the

Environment, pp 417 - 425 Schipper L and Haas R 1997 Cross-country comparisons of indicators of energy use, energy efficiency and CO_2 emissions, Energy Policy 25, pp 639-649

[3] Denholm P and Kulcinski G L 2004 Life cycle energy requirements and greenhouse gas emissions from large scale energy storage systems, Energy Convers. Manag. 45, pp 2153-2172

[4] Hondo H 2005 Life cycle GHG emission analysis of power generation systems: Japanese case, Energy 20, pp 2042–2056

[5] Pehnt M 2006 Dynamic life cycle assessment (LCA) of renewable energy technologies, Renewable Energy 31, pp 55–71

[6] Amponsah N W, Troldborg M, Kington B, Aalders I and Hough R L 2014 Greenhouse gas emissions from renewable energy sources: a review of lifecycle considerations, Renew. Sustain. Energy Rev. 39, pp 461-475

[7] Asdrubali F, Baldinelli G, D'Alessandro F and Scrucca F 2015 Life cycle assessment of electricity production from renewable energies: Review and results harmonization, Renew. Sustain. Energy Rev. 42, pp 1113-1122

[8] Raadal H L, Gagnon L, Modahl I S and Hanssen O J 2011 Life cycle greenhouse gas (GHG) emissions from the generation of wind and hydro power, Renew. Sustain. Energy Rev. 15, pp 3417-3422

[9] Turconi R, Boldrin A and Astrup T 2013 Life cycle assessment (LCA) of electricity generation technologies: overview, comparability and limitations, Renew. Sustain. Energy Rev. 28, pp 555-565

[10] CML-IA Methodology 2015, Institute of Environmental Sciences, https://www.universiteitleiden.nl/en/research/research-output/science/cml-ia-characterisation-factors

[11] SimaPro 2014, https://simapro.com/

IOP Conf. Series: Materials Science and Engineering **351** (2018) 012007 doi:10.1088/1757-899X/351/1/012007

Probabilistic framework for product design optimization and risk management

J K Keski-Rahkonen[1]

[1] Product Management Engineering, Rolls-Royce Marine, PL 220, 26101 Rauma, Finland

Abstract. Probabilistic methods have gradually gained ground within engineering practices but currently it is still the industry standard to use deterministic safety margin approaches to dimensioning components and qualitative methods to manage product risks. These methods are suitable for baseline design work but quantitative risk management and product reliability optimization require more advanced predictive approaches. Ample research has been published on how to predict failure probabilities for mechanical components and furthermore to optimize reliability through life cycle cost analysis. This paper reviews the literature for existing methods and tries to harness their best features and simplify the process to be applicable in practical engineering work. Recommended process applies Monte Carlo method on top of load-resistance models to estimate failure probabilities. Furthermore, it adds on existing literature by introducing a practical framework to use probabilistic models in quantitative risk management and product life cycle costs optimization. The main focus is on mechanical failure modes due to the well-developed methods used to predict these types of failures. However, the same framework can be applied on any type of failure mode as long as predictive models can be developed.

1. Introduction

In engineering it still appears to be an industry standard to handle product design with deterministic models and risk management with qualitative tools [1], [2], [3], [4]. This is rather surprising considering quantitative probabilistic methods are widely in use in many other fields such as finance and insurance [5]. Such methods have also been introduced to engineering applications more than 50 years ago [6]. In addition, little research exists to show the actual performance of traditional qualitative risk management methods [5], [7]. On the contrary, literature suggests that even simple quantitative models beat expert judgement at prediction and decision making within many practical applications [8] [9]. While the performance of qualitative risk management methods remains uncertain in the light of scientific research, there is evidence showing the superiority of quantitative methods [5], [10], [11]. Based on the trend seen in other fields and based on the literature on performance comparisons it appears reasonable to work towards implementing probabilistic quantitative methods in engineering practices as well. It is to be noted, however, that these methods should be simple enough to be practical.

The reliability level to which a certain product should be designed, is a common topic of debate. One party often prefers to drive manufacturing costs down, while the other argues for higher reliability. From a financial perspective, this debate boils down to a product life cycle costs optimization. In other words, there should be a reliability level that optimizes the trade-off between manufacturing and quality costs which realize in cases of failure. The idea of optimizing risk exposure in terms of costs within the design is not a new one. The literature introduces four general methods to analyse engineering decisions: Life Cycle Cost Analysis (LCCA), Utility Theory (UT), Cumulative Prospect Theory (CPT) and Life Profitability Method (LPM) [12]. UT & CPT try to incorporate the risk preferences of the true decision maker into the objective function while LCCA is based on the financial value only [12], [13], [14]. LPM is a version of LCCA, capturing more financial information by covering positive cash flows as well [12]. We sympathize with the idea of trying to incorporate cognitive biases into decision analysis tool to have the optimizer to be in line with a true human decision maker. That makes perfect sense if the purpose of the model is to mimic a human decision maker. However, we concentrate on models which are financially rational, have the potential to differ from actual human opinion and aid in making better decisions. That leaves UT and CPT out of our interest. Another argument against UT, CPT and LPM is their added layer of complexity compared to LCCA. LCCA methods are simpler since they do not require decision maker specific parameters. Decision maker and application specific risk tolerances can be implemented simply as an optimization constraint while keeping the optimization process itself as a constant.

Several LCCA approaches have been suggested [15], [16], [17], [18], [19], [20]. Sørensen goes on to classify cost optimizing types of approaches as category IV reliability methods [21]. This paper introduces an alternative LCCA framework, which pursues a simpler and more practical approach to tackle the reliability optimization in everyday engineering work, while not sacrificing too much prediction accuracy. In order to define the process, we apply well-established engineering practices but also borrow from probability theory, financial models and decision analysis.

This paper provides a framework on which to define risk levels while striving to maximize product cost-effectiveness. This is achieved by applying optimization techniques on financial risk exposure, while considering design and legal requirements as well as environmental and safety risks as constraints. First, the process is performed on component level and then the component results are extended to the system level. The introduced probabilistic design process can be expressed as the steps shown in table 1. The order of chapters on this paper also follows the step numbers in table 1, each chapter delving deeper into these steps. In addition, chapter 7 applies the framework into practice presenting a practical example.

Table 1. Process steps for probabilistic design framework.

Step number	Description of a process step
1	Apply physical or statistical model of the system to predict the probability of failure for component design options
2	Calculate expected life cycle costs for component design options
3	Consider risk management and other constraints to define acceptable design space
4	Choose feasible design with acceptable level of risk and the lowest expected life cycle costs
5	Repeat the process for all the components of the system and finally verify risks on the system level

2. Model of the system

In order to estimate the failure probability of a component, a mathematical model is required. In principle, one can work with failure data to create a statistical failure probability model. However, in many practical applications product volume is too low to derive accurate failure probability estimations for the high reliability components within reasonable period of time. Another downside with the failure statistics is that they cannot predict the reliability of a new significantly different components. Therefore, a physics-based model capable of predicting the failure probability of any design is preferable. Material mechanical failure models are well-developed and quite accurate, therefore we concentrate on them within the scope of this paper. Moreover, if components are under static loading, models can be very simple. However, for many practical applications failure can also be fatigue related. In principle, any fatigue model capable of estimating fatigue life will do. Nonetheless, the more sophisticated the model the better [22].

Fatigue models are commonly applied in a deterministic manner [4]. That is, an adequate safety margin is required against failure. The safety margin is required because the model and its inputs are not perfectly accurate. Method is valid but, a difficult question arises: what is an adequate safety margin?

Some fatigue standards recommend a minimum allowed safety margin but in practice the accuracy of the inputs is case specific. For example, if the load is known precisely, a lower safety margin against material failure is adequate compared to a scenario where the load is highly uncertain. Deterministic models have no direct way to deal with uncertainty and the indirect way of compensating for uncertainty with a safety margin remains highly subjective.

It is actually the level of uncertainty and the magnitude of safety margin that in combination define the probability of failure. There is a simple solution to dealing with input uncertainty and quantifying failure probability. We can take any commonly applied mathematical material strength model and apply it stochastically instead of using a common deterministic engineering practice. Applying the Monte Carlo method is a practical way to implement the transition to stochastic modelling but requires the sources of uncertainty to be quantified. Table 2 lists the main sources of uncertainty for mechanical failure modes and categorizes them as defined by the Joint Committee of Structural Safety (JCSS) [15]. Based on table 2 many of the listed uncertainty sources are, at least to some extent, related to statistical or model uncertainty, which can be reduced by acquiring more measurement data [4]. Common designation for statistical and model uncertainties is epistemic uncertainty [4]. Epistemic uncertainties can be considered subjective in nature, since they are related to the amount of knowledge or data one possesses. Natural uncertainties, also called as aleatory uncertainties, are objective since they are inherent in nature [4].

Working with completely or partially subjective uncertainty estimates, however, requires one to lean towards a Bayesian view of probabilities. A frequentist can only handle uncertainties based on data, which of course should be the target anyway. In the absence of measurement data, engineering decisions still need to be made. A subjective quantitative estimate of parameter uncertainty remains a more effective option than facing the problem indirectly with deterministic safety margin methods. In the real world, in the presence of epistemic uncertainties, a failure probability estimate is always tied to the knowledge of the designer and the data available, rather than being a factual property of a certain design. We can only pursue objectivity by applying more accurate models, measuring input parameters more rigorously and constantly calibrating our estimates based on field data.

Table 2. Sources of uncertainty.

Description of uncertainty	Uncertainty category
Material strength	Natural, statistical
Geometry	Natural, statistical
Load magnitude and number of cycles	Natural, statistical
Failure point simulation	Model

Acknowledging these facts, however, does not diminish the practical applicability of the approach presented in this paper. From a decision analysis point of view, the source of uncertainty is irrelevant for decision making. All the uncertainties are treated mathematically in a similar manner in any case. [15], [22], [23]

Categorization of risk sources is, nonetheless, important since the higher the contribution of natural variability to total uncertainty, the closer to true failure frequency we can expect our failure probability prediction to land. In other words, if the contribution of model and statistical uncertainties is significant, model predictions are not necessarily in line with the measured failure frequency. This fact highlights the importance of the measurements in minimizing statistical uncertainties. Another important step to tackle the issue is to apply a calibration loop between the model predictions and the number of experienced failures. Bayesian mathematics can be applied to gradually adjust the model output estimate of failure probability (prior) based on actual failure frequency to always have the best estimate (posterior) in the light of available knowledge. Whilst that is beyond the scope of this paper, see JCSS for an introduction into the topic [16].

It is common practice to measure material strength standard deviation by tensile testing. This is quite expensive and usually some level of statistical uncertainty remains on top of natural strength deviation inherent in all materials. Statistical measurement uncertainty should be taken into account while not relying completely on potentially inaccurate standard deviation provided by tests. While omitted here, it can be shown that uncertainty of uncertainty is mathematically nothing more than a larger uncertainty. In the light of this information, a lack of measurements does not prevent us from getting an estimate of the failure probability. In the absence of measurement data, input value uncertainties are naturally only higher. Probabilistic methods inherently account for parameter uncertainty, which is always case specific.

Geometry variations are defined by tolerances, therefore they can be directly controlled by design. Load uncertainty can be measured by the instrumentation of products or their operating environment. Different operating conditions and operators introduce natural load variability. In addition, naturally some statistical uncertainty remains due to a limited amount of measurement data available. Model uncertainty can be significant as well. Model uncertainty is a residual term that explains the remaining prediction inaccuracy in a component life when all the uncertainties of the model inputs are considered. Figure 1 presents the relationship between safety margin and probability of failure. Only material and load uncertainties are shown in figure 1 and they are represented by Gaussian distributions for the sake of simplicity. Safety margin is defined as a ratio between the most likely values of the material strength and the load. That is true with an assumption that the mathematical model is applied in such a manner that all the inputs are given as expected values and the model is unbiased and calibrated to give the best estimate of failure point of the material [4]. On these conditions, the failure probability can be defined as the area in which these two distributions overlap.

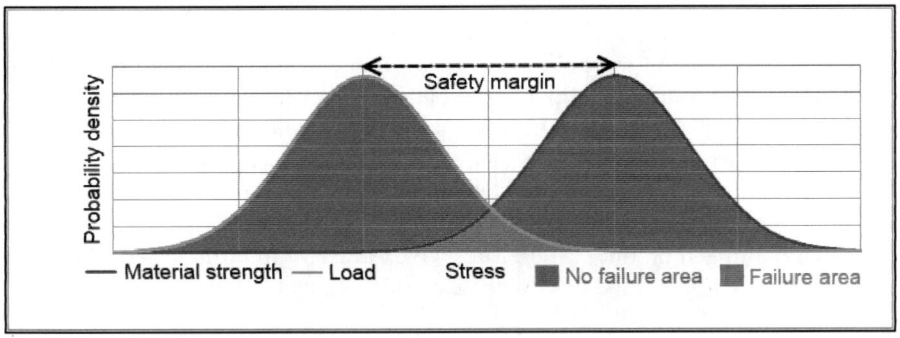

Figure 1. Safety margin and probability of failure.

There is a number of ways to solve this academic problem but we prefer a general method that allows us to define probability distributions freely to represent available measurement data as precisely as possible. We recommend to apply the Monte Carlo method on top of a fatigue model in order to estimate the probability of failure. Figure 2 shows the principle behind the Monte Carlo method simulating the failure probability. In its simplest form, we take all the internal safety margin away from the underlying mathematical model or introduce a correction term to have it estimate the expected failure point [4]. This of course requires deep understanding of the model and the simulated phenomena or a significant amount of data which to use in model calibration. The Monte Carlo process varies input parameters within their probability distributions randomly. The method gives an approximation of failure probability as a ratio between unfavorable outcomes to the total number of iterations. The code for the Monte Carlo method is simple to program and implement on top of any mathematical strength model the analyst is familiar with. Therefore, it is a perfect fit for various practical engineering problems.

Engineers generally apply finite element method (FEM) to solve the stress of structures. The use of FEM is computationally demanding and therefore to reduce computation time it is preferable to seek for procedures where the Monte Carlo iteration loop is a post-processing step for FEM. Such a procedure can be easily implemented if linear FEM is applied as shown in figure 2. That is, the stress results can be scaled and the superposition principle can be applied to them. Fortunately, practical strength related problems can be often solved with reasonable accuracy by applying linear FEM.

We generally apply FEM to solve the stress distribution on the structure and post-processes the results with the Monte Carlo algorithm to solve the probability of failure. Findley criterion is generally a good choice as a multiaxial fatigue failure criterion for steel materials [24], [25]. However, if the Findley criterion is applied on a massive steel structure, the method inherently contains some internal safety margin against complete breaking of the structure. That is because the criterion does not differentiate between crack initiation and growth phases. Existence of the internal safety margin also depends on the definition of a failure: complete breakage versus the point where significant cracks first appear.

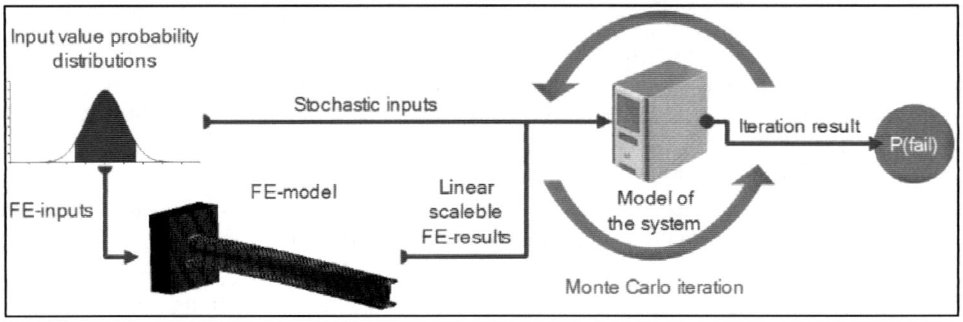

Figure 2. Process flow chart on solving failure probability.

3. Life cycle costs

Capability to solve the failure probability of a component has applications not only on risk management but also on minimizing product life cycle costs. We define life cycle costs as a sum of all the costs experienced during the product lifetime, including manufacturing costs. All the life cycle costs are also discounted to the same time period to be comparable. In efficient markets the product with lower life cycle costs will have a competitive edge against the product with higher life cycle costs, all else being equal. In practice, the competitive edge should realize as a higher sales margin or increased sales of an optimized product. Therefore, the exercise of minimizing life cycle costs can also be seen as a process of maximizing the customer value of a product having certain given features.

3.1. Expected manufacturing costs

By manufacturing costs, we refer to all the costs such as material and work that are due to the manufacturing and design process of a component. Chapter 2 introduced a general methodology to solve failure probability. It is also a straightforward practice to estimate manufacturing costs for a component design based on historical data of manufacturing similar components. Estimation of the two quantities for several design options allows to curve fit a function for expected manufacturing costs $E[M(f)]$. In practice, the function can be discrete but for optimization purposes continuous functions generally behave better. Of course, it is possible to optimize with a continuous function and deal with the discretization at the later phase. Therefore, a simple curve fitting will suffice for all practical purposes. Figure 3 shows an example of such a function fitted by three design points.

When varying any single design detail, $E[M(f)]$ will in many practical cases have a monotonic downward sloping shape. That is, manufacturing costs decrease while the probability of failure increases. This behavior is obvious when considering design features such as the number of bolts on a joint, plate thickness, weld a-dimension, tighter geometry tolerances, etc. Obviously, there is no reason to consider design details that increase reliability but have no adverse effect on costs. These are no-brainers and optimization techniques are not required. The method should be applied to aid on decision making in the design details that are trade-offs between reliability and manufacturing costs.

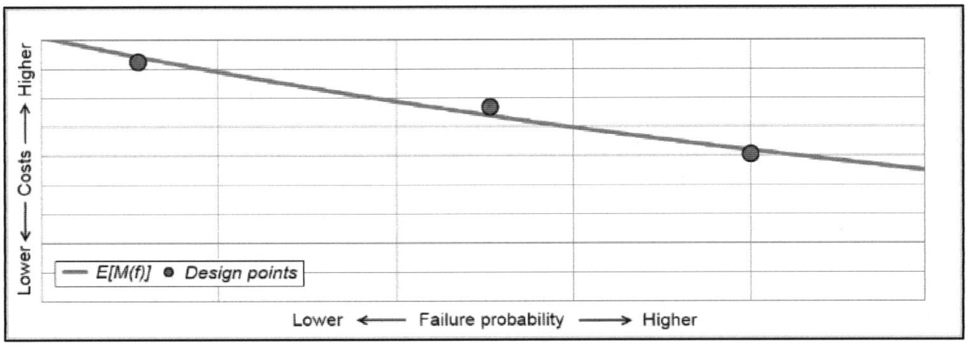

Figure 3. Estimating manufacturing costs function by curve fitting.

3.2. Expected quality costs

In this paper we define quality costs to include all the costs realizing in case of a component failure. These costs include, but are not necessarily limited to repair costs, downtime costs and costs due to brand loss. Gardoni et.al. suggest taking revaluation of an asset and optional insurance coverage into account in objective function. [12]. However, since the use of our method is limited to aiding design related engineering decisions we omit these parameters. On the other hand, Gardoni et.al. proposes a parameter to cover downtime or a revenue loss [12]. Here the optimal decision may depend on the point of view of a decision maker. The owner of the product is interested in the downtime costs while for the manufacturer the brand loss (decrease of future sales due to a bad customer experience) can be more important. In general the cost of repairing a failed component is a function of working time and spare parts. If the failure leads to downtime of a product or a brand loss for the manufacturer, resulting costs can be significant and need to be factored in. Quality costs in general are highly application specific but it should be possible to establish a reasonable estimate if some effort is put into it. Estimating quality costs is nothing more than a quantitative expression of the financial impact of failure, an exercise that is often performed qualitatively for risk matrix definition [26].

Quality costs Q_i only realize in case of failure, therefore we should be interested in expected quality costs which are a product of the probability of failure and the cost of failure. The probability of failure as a quantity needs to be tied to a time period. For many products, one annum is a natural choice for the length of the time period. Annual probability of failure f_i does not need to be constant for every period i, rather it may increase towards the end of product lifetime due to the cumulative nature of fatigue damage. For the sake of simplicity graphs in chapter 3 are plotted as if the annual probability of failure were a constant quantity.

If manufacturing costs occur at time t_0 then quality costs occur at time t_{0+i}, where i is an arbitrary time period number. Therefore, to compare apples to apples, we need to discount the expected quality costs from every failure period to the same time period as expected manufacturing costs. This is a standard practice in finance and can be carried out by applying the discounted cash flow formula [27]

$$E[Q(f)] = \sum_{i=1}^{n} \frac{f_i * Q_i}{(1+r)^i} \qquad (1)$$

for summing the quality cost cash flows over the product lifetime n. From manufacturer's perspective, quantity n on the formula (1) can be considered as warranty time as well. The discount term r is usually an input from the financial department. In general discount term r can be defined as a cost of capital plus stock market real risk premium. For private companies that can be approximated by summing long term credit interest rate and earnings yield. The expected quality costs function *E[Q(f)]* has an exponentially increasing shape as the probability of failure increases as shown in figure 4. This is in line with intuition, when the probability of failure approaches zero, quality costs should approach zero as well.

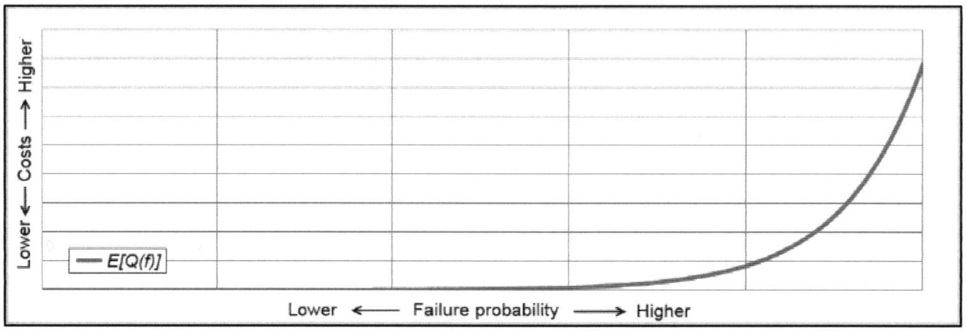

Figure 4. Quality costs function.

3.3. Expected life cycle costs

Following Gardoni et.al. we define the expected life cycle costs *E[L(f)]* as a sum of expected manufacturing cost, discounted expected quality costs and discounted expected life maintenance costs *E[C]* according to formula [12]

$$E[L(f)] = E[Q(f)] + E[M(f)] + E[C]. \tag{2}$$

Since *E[C]* is not a function of failure probability *f*, it can be set to zero while having no impact on the optimization result in terms of f [20]. *E[L(f)]* has a parabolic shape as shown in figure 5. The minimum of this parabola is an optimum reliability level, considering manufacturing and quality costs. On a free optimization case this is the optimum reliability level to target in the design.

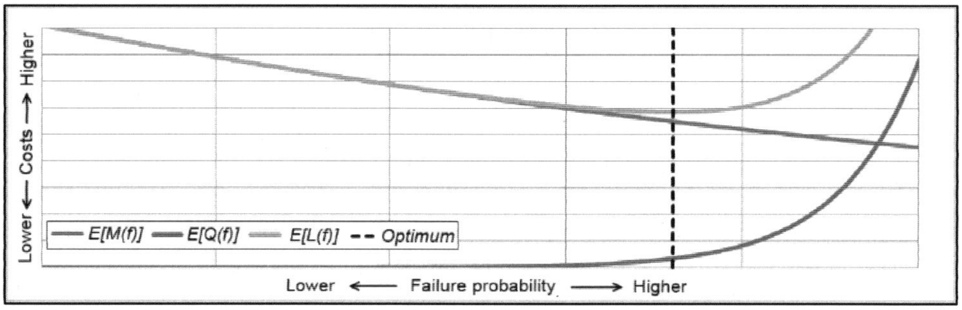

Figure 5. Life cycle costs function.

The optimum is shown in figure 5 as a vertical dashed line. Of course, in many practical engineering applications it is not possible to have a perfectly accurate estimate of *E[L(f)]*. Fortunately, practical experience shows that the optimal reliability level is not highly sensitive to the accuracy of the underlying parameters. That is, the method has been found to be a useful decision analysis tool in practice even in the presence of significant uncertainties. A recommended practice is to run a sensitivity analysis or the Monte Carlo simulation to quantify the confidence on reliability optimization as well. However, within the scope of this paper we treat reliability optimization in a deterministic manner.

4. Risk management and other constraints

As always in case of a practical optimization problem, constraints should be identified. Risk management can introduce a set of constraints to the design process. In addition, any rules and design guidelines that are considered important, can be introduced as optimization constraints. This is true even if the constraints are not a function of failure probability or directly related to the objective function i.e. life cycle costs function [15]. Quite often, there are legal constraints such as standards that need to be followed or classification society rules that need to be fulfilled. Taking the constraints into account is straightforward. Just limit feasible design space by excluding the design options that are outside of the chosen constraints.

There are several approaches to deal with risk but we introduce only the four more common ones. The first and the simplest is a risk neutral approach. The risk neutral decision maker does not care about risk, but only tries to maximize return [12]. In our application that would be equal to minimizing life cycle costs. In theory this is a valid strategy but not commonly applied in practice due risk aversion typical for most decision makers [9], [12], [28]. The second possible strategy is to define an acceptable level of risk and optimize only within this constraint [19]. This is a common approach for risk averse decision makers. The third option is to form a risk corrected return measure which quantifies how many units of return is achieved per unit of risk [29]. The third approach is common for risk seekers and in general in the field of finance. The fourth option is to build risk preferences into objective function [13], [12]. In this paper we choose to use the second approach due to its simplicity and conservativeness.

There are also three main categories of risks to consider when designing a product: financial, environmental and safety. Following recommendations given by JCSS, tangible and intangible risks are clearly separated [15]. Tangible i.e. financial risks are part of the reliability level optimization introduced in chapter 3. Therefore, depending on the risk profile, one may not want to add additional risk management criteria for financial risks. It is possible, at least qualitatively, to estimate consequences of failure to environment and safety of the people involved. Yet it is suggested to quantify intangible risks too whenever possible, since that allows the exploitation of generally accepted risk levels [30], [31]. Estimation of consequences of failure enables the definition of tolerable component failure probabilities. These tolerable risk management limits work as a constraint on the life cycle costs optimization.

As a practical note, we consider design requirements such as manufacturability, stiffness, vibrations, stability, noise, flow characteristics and classification society rules as deterministic constraints for the optimization problem. Risk management is naturally another layer of constraints.

5. Optimal component design

Chapter 2 gave a framework to predict failure probabilities for design options, chapter 3 showed how to estimate life cycle costs function and finally chapter 4 introduced risk management constraints for the design space. Figure 6 sums it all up in one graph. The feasible design space is shown on the white background after considering shaded area of constraints.

From this point on it is a simple exercise to minimize the *E[L(f)]* function to find a global optimum for the reliability level (1-*f*). In figure 6 the vertical dashed line, which represents the optimum reliability level, is now shifted towards left due to the chosen constraints. We also chose to run the optimization with a continuous function but the actual design options are often discrete. In that case, naturally the closest feasible design option to the optimum should be chosen.

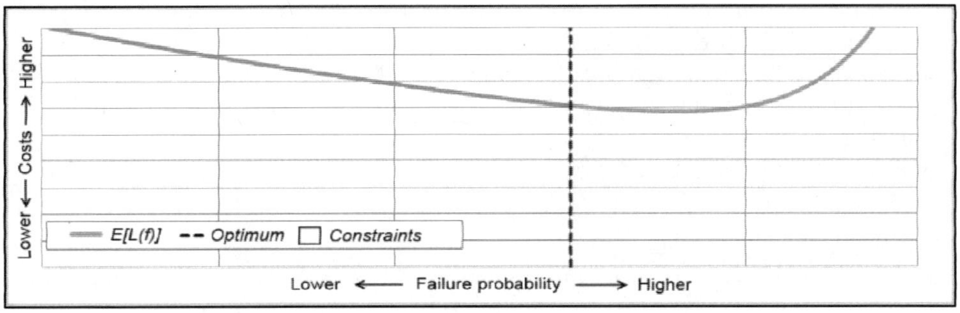

Figure 6. Life cycle costs function with constraints.

6. System level

Chapters 3-5 introduced a framework to optimize on a component level. Component level life cycle cost optimization should be adequate for the product as a whole as well. This is not necessarily so from the risk perspective. If the failure probabilities of individual components are less than perfectly correlated, the failure probability of a system will always be higher than the highest individual component failure probability. Therefore, for risk management reasons it is important to consider the system level failure probability as well.

If we let all the component level failure probabilities f_k be independent, the system level probability of failure f_s boils down to a simple sum product equation [32]

$$f_s = 1 - \prod_{k=1}^{x}(1 - f_k), \tag{3}$$

x being the number of components. Equation (3) can also be applied if one prefers a simple and conservative approach to risk management. In practice, component level failure probabilities are usually correlated. In the case of mechanical failure modes, the whole system often experiences the same load. If different components are dimensioned with the same failure criteria, model uncertainty may bias component failure probabilities in a mutual direction. On the other hand, the natural part of the material strength uncertainty is fundamentally uncorrelated. That is because material strength variation is mainly a local phenomenon due to impurities and micro cracks. Therefore, case specific component level correlation coefficients can be estimated with a reasonable accuracy based on quantified sources of uncertainty.

On the other end of the spectrum, in a perfectly correlated system, the product level failure probability equals to the highest component failure probability. The reality is usually between the two extremes, even negative correlation coefficients are conceivable. Thoft-Christensen et.al. present an analytical equation to solve the system level probability of failure for any correlation coefficient [33]. In practical engineering applications, the system level probability of failure for less than perfectly correlated components, can also be solved with the Monte Carlo algorithm.

7. Application example

This chapter applies the introduced framework into a simple example. Let us consider a high strength steel rod with a solid circular cross-section under fully reversed constant amplitude uniaxial uniform stress. The rod is a part of system that has the design life of 10 years. The goal is to optimize the rod diameter in terms of life cycle costs within the risk management constraints. The structure is assumed to be completely fatigue strength limited having no other limiting failure modes such as buckling or vibrations. To avoid separating crack initiation and growth phases, we define the failure as a point where significant cracks first appear. For this simplified example we also neglect fatigue strength correction factors such as size and surface quality. With these assumptions we can directly compare stress amplitude to a SN-curve derived from fatigue tensile tests. Moreover, the failure criteria is $\sigma_S > \sigma_A$, where σ_S is the fatigue strength taken from the SN-curve and σ_A is the stress amplitude in the plate. Therefore, the safety factor is defined simply as $S = \frac{\sigma_A}{\sigma_S}$. Chosen failure model is unbiased but we assume some model uncertainty in terms of failure point estimation. We apply Matlab based Monte Carlo code with one million iterations following the process shown in figure 2 to solve the failure probability estimates. The stress amplitude and the material fatigue strength are assumed to be uncertain but constant over the design life. The annual number of cycles is uncertain and independent for every year over the design life. FEM is not required in this case since the stress amplitude is simply defined by formula

$$\sigma_A = \frac{4*F_a}{\pi*D^2}. \tag{4}$$

Table 3 lists the required input parameters for this probabilistic product design optimization problem. We use two types of probability distributions for input parameters; Gaussian distribution and constant probability distribution. Applied SN-curve is a linear interpolation between σ_{S1e2} and σ_{S1e7} in the log-

log scale. We assume that component manufacturing costs consists completely of material costs. Cost of failure is also known and constant over time. The expected life cycle cost function *E[L(f)]* is calculated separately for low and high uncertainty environments (HUE and LUE). These two environments can be seen as scenarios of different level of available measurement data. The expected life maintenance costs *E[C]* are neglected in a similar manner as explained in chapter 3.

Table 3. Input parameters for the application example.

Quantity	Symbol	Expected value	HUE deviation	LUE deviation
Fatigue strength 1e7	σ_{S1e7}	500MPa	SD 50MPa	SD 25MPa
Fatigue strength 1e2	σ_{S1e2}	1 000MPa	SD 50MPa	SD 25MPa
Load amplitude	F_a	2 000kN	SD 500kN	SD 100kN
Annual number of cycles	N	5e5	± 4e5	± 1e5
Model uncertainty	M	-	± 50MPa	± 10MPa
Discount term	r	7%	0	0
Material density	ρ	7850kg/m^2	0	0
Rod diameter	D	75-110mm	0	0
Rod length	L	1000mm	0	0
Material cost	C	10€/kg	0	0
Cost of failure	Q	1 0000€	0	0
P(f) constraint	$P(f)a$	1%	0	0

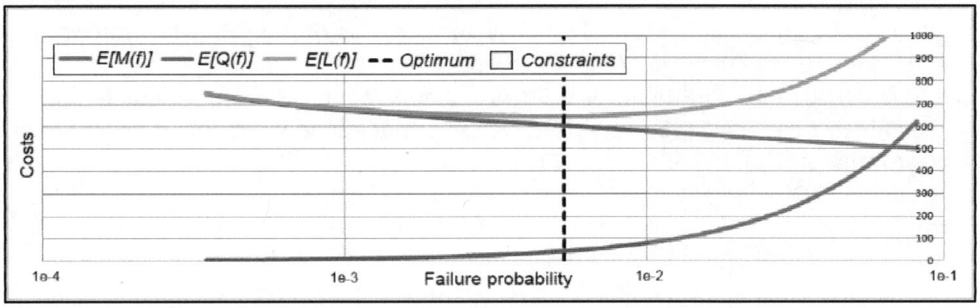

Figure 7. Life cycle costs function for HUE.

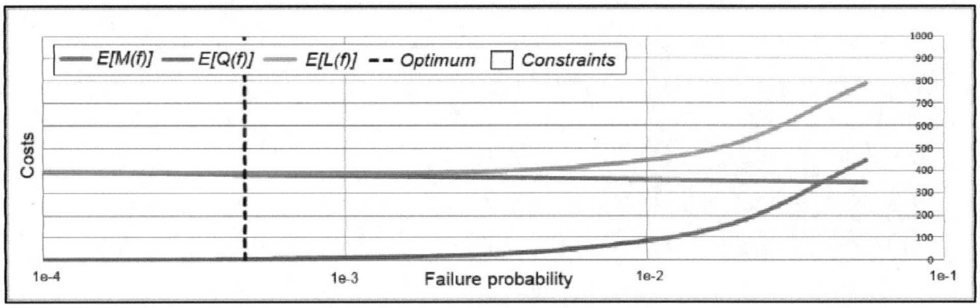

Figure 8. Life cycle costs function for LUE.

Figures 7 and 8 show optimization space for HUE and LUE. As expected *E[M(f)]* has a downward slope, *E[Q(f)]* has an upward slope while *E[L(f)]* as a sum of the two has a parabolic shape. The Matlab code to calculate presented results is shown in the appendix. The quality costs are discounted to present value according to equation (1) with an annual time step.

IOP Conf. Series: Materials Science and Engineering **351** (2018) 012007 doi:10.1088/1757-899X/351/1/012007

The failure probability on x-axis represents the cumulative probability over 10 year design life. The reliability constraint does not limit the optimization problem in either of the cases. In case of LUE it is obviously more cost effective to design for the higher reliability level compared to HUE. LUE also gives significantly lower expected life cycle costs in the optimum reliability level compared to HUE case. This difference is an indication of the financial value of having more data in the design phase. The optimal rod diameters are 99 mm and 79 mm for HUE and LUE respectively. Furthermore, the optimal design safety factor for HUE is 2.0, while it is only 1.3 for LUE. Clearly, the conventional method of dimensioning against the safety factor does not properly capture the uncertainty, the risk level nor the cost aspects of the design. On the other hand, the probabilistic framework requires a bit more engineering effort. In the cases of expensive structures or high product volumes the cost of additional engineering work is clearly exceeded by the cost benefits of the optimal design.

Many structural engineering problems are of course significantly more complex. FEM is required to solve the stress level due to complicated geometry. There are usually many load cases and the failure criterion needs to account for multiaxial stress state. Also, significant effort is often required to estimate the expected quality costs. Although the example above is highly simplified, the approach itself is quite general and can be applied to the assessment of more complicated engineering problems as well.

8. Conclusion

Based on our literature review, quantitative probabilistic decision analysis methods appear to outperform standard deterministic and qualitative methods. We introduced a new quantitative probabilistic framework for designing safe and more cost-effective products. The presented approach is a significant upgrade on the commonly applied safety margin methods. The method is at its strongest as a decision making tool applied on design details that are trade-offs between reliability and manufacturing costs. There are other approaches that pursue the same goal, but the one presented here is shaped towards simplicity and practicality on actual engineering work. The framework has also practical applications beyond the ones presented in this paper. One of these applications is a process of quantifying added financial value of measurements. Another interesting use case is the exercise of estimating the value of extended warranty time. These applications were not considered here but are reserved for the future work.

9. Appendix

```
%MATLAB CODE TO SOLVE FAILURE PROBABILITY AND LIFE CYCLE COSTS FOR HUE

%INPUTS
Ni = 1e3; %number of iterations
Ny = 10; %number of years
C = 10; %material costs €/kg
rho = 7850; %density kg/m^3
Qf = 10000; %cost of failure kg/m^3
r = 0.07; %discount term
Lr = 1000; %length [mm]
Dmin = 90; %diameter min[mm]
Dmax = 110; %diameter max[mm]
Dstep = 1; %diameter step[mm]
Favg = 2000000; %AVG load [N]
Fsd =   500000; %SD load [N]
Navg = 5e5; %AVG annual cycles
Ndev = 4e5; %+/- DEV annual cycles constant distribution deviation
FATsd = 50; %SD fatigue strength [MPa]
Mdev = 50; %model deviation %+/- constant distribution deviation [MPa]
Nsn = [1e2 1e7]; %SN-curve cycles
Ssn =[1000 500]; %[MPa] SN-curve strength

%cycles distribution
Nmin = Navg-Ndev;
Nmax = Navg+Ndev;

%SN curve function
SN = @(x)  10 ^ ((log10(Ssn(1)) - log10(Ssn(2))) / (log10(Nsn(1)) - log10(Nsn(2))) * (log10(x) - log10(Nsn(2))) + log10(Ssn(2)));

%iterate over all diameters
Dnum = 0;
for D = Dmin:Dstep:Dmax
Dnum = Dnum + 1;
F(Dnum, 1) = D;
F(Dnum, 2:end) = 0;
A = pi*(D/2)^2; %stress area
M(Dnum) = C * rho * A/1e6 * Lr/1000; %material costs
```

```
%interpolate fatigue strength based on SN-curve
FAT = SN(Navg*Ny);
SfDEt(Dnum) = FAT / (Favg / A); %safety factor

  %initial values of F
  for year = 1:1:Ny
  F(Dnum, year+1) = 0;
  end

  %monte carlo iteration
  for MC = 1:1:Ni
  Ntot = 0;
  Merr = (Mdev * 2) * rand(1) - Mdev;

  %Stress
  Savg = Favg / A;
  Ssd = Fsd / A;
  S = randn(1) * Ssd + Savg;

  %random component of FAT strength
  FATrand = randn(1) * FATsd;

    for year = 1:1:Ny

    %FAT strength for the year
    Na = (Nmax - Nmin) * rand(1) + Nmin;
    Ntot = Ntot + Na;
    FAT = SN(Ntot) + FATrand + Merr;

      %count the number of failures
      Sf = FAT / S;
      if Sf < 1
      F(Dnum, year+1) = F(Dnum, year+1) + 1;
              for year2 = year+1:1:Ny
              F(Dnum, year2+1) = F(Dnum, year2+1) + 1;
              end
      year = Ny;
      end

      %calculate quality costs
      if year > 1
          Q(Dnum) = Q(Dnum) + (Qf * (F(Dnum, year)/Ni-F(Dnum, year-1)/Ni)) / (1+r)^year;
      else
          Q(Dnum) = Qf * F(Dnum, year)/Ni / (1+r)^year;
      end

    end
  L(Dnum) = M(Dnum) + Q(Dnum); %life cycle costs
  end

%failure probability function
F(Dnum, 2:end) = F(Dnum, 2:end) / Ni;
Pf(Dnum) = F(Dnum, Ny);
end

semilogx(Pf,M,Pf,Q,Pf,L) %plot functions
```

10. References

[1] DNVGL-CG-0138, *Direct strength analysis of hull structures in passenger ships,* 2016.

[2] J. Oehmen, M. Ben-Daya, W. Seering and M. Al-Salamah, "Risk Management in Product Design: Current State, Conceptual Model and Future Research," in *ASME 2010 International Design Engineering Technical Conference*, Montreal, 2010.

[3] R. Jerrard, R. N. Barnes and A. Reid, "Design, risk and new product development in five small creative companies," *International Journal of Design,* pp. 21-30, 2008.

[4] P. Gardoni, A. Der Kiureghian and K. Mosalam, "Probabilistic capacity models and fragility estimates for reinforced concrete columns based on experimental observations," *Journal of Engineering Mechanics,* vol. 128, no. 10, pp. 1024-38, 2002.

[5] D. W. Hubbard, The Failure of Risk Management: Why It's Broken and How to Fix It, Wiley, 2009.

[6] A. M. Freudenthal, J. M. Garrelts and M. Shinozuka, "The Analysis of Structural Safety," *Journal of the Structural Division,* pp. 267-325, 1966.

[7] L. A. Cox, "What's wrong with risk matrices?," *Risk Analysis,* p. 497–512, 2008.

[8] A. Tversky and D. Kahneman, "Judgement under Uncertainty: Heuristics and Biases," *Science,* vol. 185, pp. 1124-1131, 1974.

[9] D. Kahneman, Thinking, Fast and Slow, Farrar, Straus and Giroux, 2011.

[10] F. Macmillan, *Risk, Uncertainty and Investment Decision-Making in the Upstream Oil and Gas Industry,* Aberdeen, 2000.

[11] L. B. Chincarini, *A Comparison of Quantitative and Qualitative Hedge Funds,* San Francisco, 2010.

[12] P. Gardoni, F. Guevara-Lopez and A. Contento, "The Life Profitability Method (LPM): A financial approach to engineering decisions," *Structural Safety,* vol. 63, pp. 11-20, 2016.

[13] D. Kahneman and A. Tversky, "Prospect Theory: An Analysis of Decision under Risk," *Econometrica,* vol. 47, pp. 263-291, 1979.

[14] J. v. Neumann, O. Morgenstern, H. W. Kuhn and A. Rubinstein, Theory of games and economic Behavior, Princeton University Press, 2007.

[15] Joint Committee of Structural Safety, "Risk Assessment in Engineering," *Principles, System Representation & Risk Criteria,* 2008.

[16] Joint Committee of Structural Safety, "Reliability Based Code Calibration," 2002.

[17] J. S. Nathwani, N. C. Lind and M. D. Pandey, "Affordable Safety By Choice: The Life Quality Method," 1997.

[18] Y. K. Wen and Y. J. Kang, "Minimum building life-cycle cost design criteria I methodology and II applications," *Journal of Structural Engineering,* vol. 127, no. 3, 2001.

[19] D. M. Frangopol, K.-Y. Lin and A. C. Estes, "Life-Cycle Cost Design of Deteriorating Structures," *Journal of Structural Engineering,* vol. 123, no. 10, p. 1390–401, 1997.

[20] R. Kumar, P. Gardoni and M. Sanches-Silva, "Effect of cumulative seismic damage and corrosion on life-cycle cost of reinforced concrete bridges," *Journal of Earthquake Engineering & Structural Dynamics,* vol. 38, no. 7, p. 887–905, 2009.

[21] J. D. Sørensen, Notes in Structural Reliability Theory And Risk Analysis, Aalborg, 2004.

[22] Joint Committee of Structural Safety, Interpretation of Uncertainties and Probabilities in Civil Engineering Decision Analysis, 2008.

[23] Federal Institute of Materials Research and Testing, F08a Guideline for the Assessment of Existing Structures, Berlin, 2006.

[24] R. Rabb, Todennäköisyysteoriaan pohjautuva väsymisanalyysi, Books on Demand, 2014.

[25] R. Roger , C. Lönnqvist and J. Kaas, "Multiaxial Fatigue Criteria Applied to Medium Speed Diesel Engines," in *ICMFF9,* Parma, 2010.

[26] N. J. Duijm, "Recommendations on the use and design of risk matrices," *Safety Science,* vol. 76, p. 21–31, 2015.

[27] K. Bennouna, G. Meredith and T. Merchant, "Improved capital budgeting decision making: evidence from Canada," *Management Decision,* vol. 48, no. 2, p. 225–6, 2010.

[28] K. Goda and H. P. Hong, "Optimal Seismic Design Considering Risk Attitude, Societal Tolerable Risk Level, and Life Quality Criterion," *Journal of Structural Engineering,* vol. 132, no. 12, p. 2027–35, 2006.

[29] W. F. Sharpe, "Mutual Fund Performance," *Journal of Business,* pp. 119-138, 1966.

[30] Joint Committee of Structural Safety, Risk Acceptance Criteria, 2008.

[31] Eurocode 0, EN 1990, Basis of Structural Eurocodes, 2001.

[32] M. Rausand and A. Høyland, System Reliability Theory: Models, Statistical Methods, and Applications, Wiley, 2004.

[33] P. Thoft-Christensen and J. D. Sørensen, "Reliability of structural systems with correlated elements," *Applied Mathematical Modelling,* vol. 6, no. 3, pp. 171-178, 1982.

Statistical Bayesian method for reliability evaluation based on ADT data

Dawei Lu[1], Lizhi Wang[2], Yusheng Sun[1] and Xiaohong Wang[1]

[1] School of Reliability and System Engineering, Beihang University, China
[2] Unmanned System Institute, Beihang University, China

wxhong@buaa.edu.cn

Abstract. Accelerated degradation testing (ADT) is frequently conducted in the laboratory to predict the products' reliability under normal operating conditions. Two kinds of methods, degradation path models and stochastic process models, are utilized to analyze degradation data and the latter one is the most popular method. However, some limitations like imprecise solution process and estimation result of degradation ratio still exist, which may affect the accuracy of the acceleration model and the extrapolation value. Moreover, the conducted solution of this problem, Bayesian method, lose key information when unifying the degradation data. In this paper, a new data processing and parameter inference method based on Bayesian method is proposed to handle degradation data and solve the problems above. First, Wiener process and acceleration model is chosen; Second, the initial values of degradation model and parameters of prior and posterior distribution under each level is calculated with updating and iteration of estimation values; Third, the lifetime and reliability values are estimated on the basis of the estimation parameters; Finally, a case study is provided to demonstrate the validity of the proposed method. The results illustrate that the proposed method is quite effective and accuracy in estimating the lifetime and reliability of a product.

1. Introduction

For today's highly reliable products, it is difficult to estimate their field reliabilities due to the extremely long test time under field conditions. Therefore, ADT is frequently conducted in the laboratory for predicting the products' reliability under working conditions. The main idea of ADT is to test products under harsh environmental stresses to accelerate the performance degradation on the condition of constant failure mechanism and obtain sufficient data for reliability analysis in a short time, lower costs and higher efficiency. So it has many applications, such as the reliability analyses of liquid-crystal display (LCD) [1], light-emitting diode (LED) [2] and batteries [3][4].

The core method of ADT-based reliability estimation is to find a suitable degradation model which is capable of incorporating the relationships between the applied stresses and the product's degradation [5]. The existing degradation analysis methods can be categorized into two broad classes [6], degradation path models and stochastic process models which are widely used in different combination and some extension.

Due to the stochastic process models can describe the temporal variation of the degradation process in a finite time interval, they have gained more attention from researchers, especially the typical ones like Wiener process, Gamma process [7][8]. Because Wiener process has a higher generalization and a strong ability to describe a nonlinear degradation process. It gets more intensive applications. Moreover, products in practice, their degradation increment in an infinitesimal time interval can be viewed as an additive superposition of a large number of small external effects, while Wiener process with its normally distributed increment can perfectly satisfy this property [6].

However, there are still some limitations need to consider when adopting Wiener process in ADT. Usually, when combining Wiener process and acceleration model to estimate the degradation process, the parameters are hard to calculate. Traditional estimation method fit degradation ratio with least square method and maximum likelihood estimation algorithm. But the solution process and the estimation result of degradation ratio are imprecise which may affect the accuracy of the acceleration model and the extrapolation value. The Bayesian method is proposed to solve this problem for its great capability of data fusion and the ability of parameter inference. In literature, Wang [9] combine Bayesian inference Gamma process with ADT to predict lifetime. Xu [10] propose a fully Bayesian approach to deduce parameters in SSALT. The data processing in these literatures always follows the same pattern which is to unify the degradation data from different stress level into the unified data at the working stress level before estimation. The accuracy of degradation data and some key information may lose during the processing which may also affect the veracity.

In this paper, a new data processing and parameter inference method based on Bayesian method is proposed to handle degradation data. The main idea of this method is to utilize the estimation values and processed data of previous stress level to contribute the estimation under next stress level. It can maximize the utilization of test information to gain the best result through update and iteration. It can also eliminate the deviation effect caused by data processing and improve the estimation accuracy. This method can be a useful tool to apply in ADT design optimization.

The remainder of this paper is organized as follows. In Section 2, general degradation model based on Wiener process is introduced, and the reliability function is deducted. In Section 3, first, the initial value of degradation model is estimated, then the parameters of prior and posterior distribution under each stress level are obtained with data processing under every stress level. Finally, the reliability estimation value is calculated on the basis of the estimation values and parameters. In section 4, a case study is provided to demonstrate the validity of the proposed parameter estimation method as well as the reliability estimation method. Some conclusions were drawn in Section 5.

2. Degradation model based on wiener process

We assume that the degradation follows a Wiener process, then the degradation is modeled by the following

$$Y(t) = \sigma B(t) + d(s) \cdot t + y_0 \tag{1}$$

Where $Y(t)$ is the performance degradation process of product, $B(t)$ is the standard Brownian motion with the mean value of 0 and the variance of t, denoted as $B(t) \sim N(0,t)$, σ is the diffusion coefficient, which is a constant and does not change with stress or time, y_0 is a known initial value of product performance, and $d(s)$ is the drift coefficient, which represents the degradation rate of product and is a function of stress. It can be used to combine degradation model and acceleration model. Suppose the acceleration model of degradation rate is given by

$$d(s) = \exp[A + B\varphi(s)] \tag{2}$$

Where $\varphi(s)$ is a given functions of stress factor s. then $d(s)$ can be calculated by factor A and B and the relationship between stress and degradation data, the acceleration model, can also be built.

Known from the property of the Wiener process, the degradation increment ΔY during the unit time Δt follows a normal distribution with the mean of $d(s) \cdot \Delta t$ and the variance of $\sigma^2 \Delta t$ i.e.,

$$\Delta Y \sim N(d(s) \cdot \Delta t, \sigma^2 \Delta t) \tag{3}$$

Let k be the failure threshold and the product fails when $Y(t)-k<0$, then the first passage time t to the threshold k has an inverse Gaussian distribution. The probability density function of the first passage time t is

$$\mathrm{f}(t; y_0, \mathrm{k}) = \frac{k - y_0}{\sigma\sqrt{2\pi t^3}} \exp\{-\frac{[(k - y_0) - d(s) \cdot t]^2}{2\sigma^2 t}\} \tag{4}$$

Then the reliability function of t is given by

$$R(t)=\Phi\left[\frac{l-y_0-d(s)t}{\sigma\sqrt{t}}\right]-\exp\left(\frac{2d(s)(l-y_0)}{\sigma^2}\right)\Phi\left[-\frac{l-y_0+d(s)t}{\sigma\sqrt{t}}\right] \tag{5}$$

From the function above, we know that the key parameters for reliability prediction are $d(s)$ and σ^2. After these parameters are obtained, the reliability prediction can be made.

3. Parameter and reliability estimation based on Bayesian method

3.1. Estimation of the initial value of A, B

We provide the method for estimating the unknown parameters in step stress accelerated degradation testing (SSADT). Suppose a set of stresses is $S=(s_1, s_2 \dots s_m)$ where m (m>2) is the number of stress levels, while the number of the degradation increment data under the each stress level is $\Delta Y_i=\left(\Delta y_{i1},\Delta y_{i2},\cdots\Delta y_{in_i}\right)$, where i= 1…m. Then the Linear Regression Equation can be easily obtained by (1) and (2).

$$E(y_i(t))=\exp(A+B\varphi(s_i))\cdot t+y_0 \tag{6}$$

$$\ln(E(\Delta y_i))=B\varphi(s_i)+A+\ln(\Delta t) \tag{7}$$

By fitting the degradation increment data Δy_i under the each stress level s_i with (7), the initial value of A, B, which is A_0 and B_0 can be obtained. In addition, when the degradation data is hard to collect, A_0 and B_0 can be estimated through the degradation history or engineer experiences and then updated with real time test data in the following test process.

3.2. Data processing under s_1

The main idea of Bayesian method is to specify some prior probability, which is then updated to a posterior probability in the light of new, relevant data. That means the estimation under the next stress level can be contributed by the previous level. So, data under stress level s_1 should be handled first.

Known from the property of (3) and conjugate prior distribution theory that the product between the drift coefficient $d(s)$ and time interval Δt follows the normal distribution and the product between the diffusion coefficient σ^2 and time interval Δt follows the inverse gamma distribution. Which is

$$d(s)\cdot\Delta t \sim N(\mu,\tau^2) \tag{8}$$

$$\sigma^2\cdot\Delta t \sim IG(b,a) \tag{9}$$

Where μ, τ are the mean value and the standard deviation of normal distribution, while a, b are the scale and shape parameter of gamma distribution.

In terms of data $\Delta Y_1=\left(\Delta y_{11},\Delta y_{12},\cdots\Delta y_{1n_1}\right)$ which is under stress level s_1, the initial value of hyper parameters in (8) and (9) can be calculated. Known from the property of conjugate prior distribution, the calculation formula is given by

$$a_1=\frac{1}{2}\sum_{j=1}^{n_1}\left(\Delta y_{1j}-\overline{\Delta Y_1}\right)^2 \tag{10}$$

$$b_1=\frac{n_1}{2} \tag{11}$$

$$\mu_1=\frac{1}{n_1}\sum_{j=1}^{n_1}\Delta y_{1j} \tag{12}$$

$$\tau_1^2=\frac{1}{n_1}\cdot\frac{1}{n_1-1}\sum_{j=1}^{n_1}\left(\Delta y_{1j}-\mu_1\right)^2 \tag{13}$$

Where $\overline{\Delta Y_1}$ the mean of data under is s_1, n_1 is the number of the data, and a_1, b_1 are the scale and shape parameter of gamma distribution (9) while μ_1, τ_1^2 are the mean value and the standard deviation of normal distribution (8) .Then known from the property of normal distribution the estimation value of $d(s_1)\cdot\Delta t$ under s_1 is

$$E(d(s_1)\cdot\Delta t)=u_1 \tag{14}$$

3.3. Prior and posterior distribution under s_2

Known from (2) and the initial value of A, B, which is A_0 and B_0, for s_i and s_j the specific value of acceleration model is

$$p_{ij}=d(s_i)/d(s_j)=\exp(-B_0\cdot(1/s_i-1/s_j)) \tag{15}$$

In terms of (8) one already know that the product between the drift coefficient and time interval under s_j follows normal distribution, i.e.

$$d(s_j)\cdot\Delta t \sim N(\mu_j,\tau_j^{\,2}) \tag{16}$$

Then on the basis of normal distribution one can obtain

$$d(s_i)\cdot\Delta t=p_{ij}\cdot d(s_j)\cdot\Delta t \sim N(p_{ij}\cdot\mu_j,p_{ij}^{\,2}\cdot\tau_j^{\,2}) \tag{17}$$

Based on the property of Wiener process that σ is the diffusion coefficient, which is a constant and does not change with stress or time, so the distribution of $\sigma^2\cdot\Delta t$ does not change as well. Then after the specific value is conducted, the hyper parameters calculated under s_1 can convert to prior distribution of parameter under s_2 with

$$a_{20}=a_1 \tag{18}$$

$$b_{20}=b_1 \tag{19}$$

$$u_{20}=p_{21}\cdot u_1 \tag{20}$$

$$\tau_{20}^2=p_{21}^2\cdot\tau_1^2 \tag{21}$$

Where μ_{20}, $\tau_{20}^{\,2}$ are the mean value and the standard deviation of the prior distribution under s_2, which is a normal distribution, while a_{20}, b_{20} are the scale and shape parameter of the prior distribution under s_2, which is a gamma distribution.

As discussed above, the product between the drift coefficient and time interval still follows the normal distribution and the product between the diffusion coefficient and time interval follows the inverse gamma distribution as well. So the hyper parameter of posterior distribution is given by

$$a_2=\alpha_{20}+\frac{n_2}{2}\left(\frac{1}{n_2}\sum_{j=1}^{n_2}\left(\Delta y_{2j}-\overline{\Delta Y_2}\right)^2\right)+\frac{n_2}{2}\frac{\left(\overline{\Delta Y_2}-p_{21}\cdot\overline{\Delta Y_1}\right)^2}{n_2/n_1+1} \tag{22}$$

$$b_2=b_{20}+\frac{n_2}{2} \tag{23}$$

Where a_2, b_2 are the scale and shape parameter of the posterior distribution under s_2, n_2 is the number of degradation data, and $\overline{\Delta Y_2}$ is its mean value.

Then the estimation value of $\sigma^2\cdot\Delta t$ is

$$E\left(\sigma^2\cdot\Delta t\,|\,\Delta y_2\right)=\frac{a_2}{b_2-1} \tag{24}$$

Next, the mean value and the standard deviation of the posterior distribution under s_2, which is μ_2, $\tau_2^{\,2}$ can be expressed as

$$u_2=\frac{\overline{\Delta y_2}/E(\sigma^2\cdot\Delta t)+u_{20}\tau_{20}^{\,-2}}{1/E(\sigma^2\cdot\Delta t)+\tau_{20}^{\,-2}} \tag{25}$$

$$\tau_2^2=\frac{1}{1/E(\sigma^2\cdot\Delta t)+\tau_{20}^{\,-2}} \tag{26}$$

So the estimation value of $d(s_2)\cdot\Delta t$ can be given by

$$E(d(s_2)\cdot\Delta t)=u_2 \tag{27}$$

3.4. Prior and posterior distribution under s_i

The Prior and posterior distribution of s_i can be calculated the same as s_2, so the solving process is established. Suppose the hyper parameters of posterior distribution under $s_{i\text{-}1}$ are $a_{i\text{-}1}$, $b_{i\text{-}1}$, $\mu_{i\text{-}1}$, $\tau_{i\text{-}1}^2$ respectively. Then A_0 and B_0 can be updated by(7), and by further calculating with (15)(18)(19)(20)(21) the scale and shape parameter of gamma distribution, α_{i0} and b_{i0}, as well as the mean value and the standard deviation of normal distribution, μ_{i0} and τ_{i0}^2, can be obtained.
Finally, the hyper parameters of posterior distribution can be given on the basis of the hyper parameters of prior distribution, which are

$$a_i = \alpha_{i0} + \frac{n_i}{2}\left(\frac{1}{n_i}\sum_{j=1}^{n_i}\left(\Delta y_{ij} - \overline{\Delta Y_i}\right)^2\right) + \frac{n_i}{2}\frac{\left(\overline{\Delta Y_i} - p_{ii-1}\cdot\overline{\Delta Y_{i-1}}\right)^2}{n_i/n_{i-1}+1} \tag{28}$$

$$b_i = b_{i0} + \frac{n_i}{2} \tag{29}$$

$$E\left(\sigma^2\cdot\Delta t \mid \Delta y_i\right) = \frac{a_i}{b_i - 1} \tag{30}$$

$$u_i = \frac{\overline{\Delta y_i}/E(\sigma^2\cdot\Delta t\mid\Delta y_i) + u_{i0}\tau_{i0}^{-2}}{1/E(\sigma^2\cdot\Delta t\mid\Delta y_i) + \tau_{i0}^{-2}} \tag{31}$$

$$\tau_i^2 = \frac{1}{1/E(\sigma^2\cdot\Delta t\mid\Delta y_i) + \tau_{i0}^{-2}} \tag{32}$$

$$E(d\left(s_i\right)\cdot\Delta t) = u_i \tag{33}$$

Where $\overline{\Delta Y_{i-1}}$, $\overline{\Delta Y_i}$ are the mean value under $s_{i\text{-}1}$, s_i, respectively while $n_{i\text{-}1}$, n_i are the number of degradation data. μ_i, τ_i^2 are the mean value and the standard deviation of the posterior distribution under s_i, while a_i , b_i are the scale and shape parameter of the posterior distribution under s_i. So, with the calculation above, the hyper parameters of posterior distribution under s_i and other parameters are fully obtained.

3.5. Reliability estimation

According to the parameter estimation method above, m estimation values of product between the drift coefficient and time interval under m stress levels are gained. It can be written as $U = (E(d\left(s_1\right)\cdot\Delta t), E(d\left(s_2\right)\cdot\Delta t), \cdots E(d\left(s_m\right)\cdot\Delta t))$. Then (2) can be transformed into

$$E(d\left(s_i\right)\cdot\Delta t) = \exp[A + B\varphi(s_i)]\cdot\Delta t \tag{34}$$

Next, from (34) the final value of *A, B* which is $\hat{A}$ and $\hat{B}$, can be obtained by fitting with the least-squares method, next the drift coefficient can be expressed by plugging working stress s_0 in (5) and the estimation value of σ^2 under the last stress level can be marked as final estimation value $\hat{\sigma}^2$. Finally, the reliability estimation value of product can be obtained.

4. Case study

In this section, we demonstrate the proposed model and the inferential method based on a numerical example. This example is used to demonstrate the validity and superiority of the proposed parameter estimation method. Suppose there is a temperature step-stress ADT applied to a product, where there are 4 test units and 4 temperature levels (60 °C, 80 °C, 100 °C, and 120 °C). The time of each level is 1250, 750, 500 and 500 h, and the inspection time interval is 5 h.

4.1. Degradation and acceleration model

As the stress factor is temperature, the Arrhenius acceleration function is applicable, so

$$\mathrm{d}(T) = \exp[A + B/T] \tag{35}$$

Where T is the absolute temperature, A and B are the parameters. Suppose the degradation follows Winner process, then known from the discussion of section 2, the degradation increment ΔY during the unit time Δt is subject to a normal distribution with the mean of $d(s)\cdot\Delta t$ and the variance of $\sigma^2\Delta t$, which is (3) and the reliability model is (5), Let the initial value of y_0 be 100 and the failure threshold l be 40.

4.2. Estimation of the initial value of A, B

Based on the degradation data gathered from each temperature level (60 °C, 80 °C, 100 °C, and 120 °C), the degradation increment data ΔY_{60}, ΔY_{80}, ΔY_{100}, ΔY_{120} are obtained, respectively. Then (7) is applied to fit the degradation increment data. So the initial value of A, B, which is A_0 and B_0 is calculated.

4.3. Prior and posterior distribution under each temperature level

4 temperature levels are applied to the product, so let 60 °C be the initial stress, the calculation sequence is 60 °C, 80 °C, 100 °C and 120 °C.

4.3.1. Data processing under 60 ℃

Known from the property of conjugate prior distribution, (8) and (9) are chosen to be the form of the prior and posterior distribution of the product between the drift coefficient and time interval and the product between the diffusion coefficient and time interval.
Then the hyper parameters calculated with (10)~(13) are shown in Table 1.

Table 1. Hyper Parameters of Distribution under 60 °C

a_1	b_1	μ_1	τ_1^2
0.3284	100	0.0239	1.6501e-5

4.3.2. Prior and posterior distribution under 80 ℃

The specific value of the drift coefficient under 80 °C and 60 °C can be known from(15), which is $p_{2,1}$=2.2361. Then by combining (18)~(21) the hyper parameters of prior distribution under 80 °C are calculated and shown in Table 2.

Table 2. The Hyper Parameters of Prior Distribution under 80 °C

a_{20}	b_{20}	μ_{20}	τ_{20}^2
0.3284	100	0.0534	8.2502e-5

Then based on degradation increment data, the hyper parameters of prior distribution and (22)(23)(25)(26), the hyper parameters of posterior distribution under 80 °C are gained and shown in Table 3.

Table 3. The Hyper Parameters of Posterior Distribution under 80 °C

a_2	b_2	μ_2	τ_2^2
0.4918	160	0.0535	8.0359e-5

4.3.3. Prior and posterior distribution under 100 ℃ and 120 ℃

As the discussion in section 3, the Prior and posterior distribution under 100 °C and 120 °C as well as the parameter estimation values under every temperature levels are known from (15), (18)~(21), (28)~(33). The estimation values are shown in Table 4.

Table 4. The Parameter Estimation Values under Every Temperature Levels

	60°C	80°C	100°C	120°C
$d(T)\cdot\Delta t$	0.0239	0.0535	0.1098	0.2112
$\sigma^2\cdot\Delta t$	0.0033	0.0031	0.0033	0.0032

4.3.4. Reliability estimation

From the estimation values in Table 4 and (34) the final value of A, B which is $\hat{A}$ and $\hat{B}$ are obtained, that is $\hat{A}$= 4824.1 and $\hat{B}$= 92274.2. Next, the drift coefficient under 25 °C is extrapolated with $\hat{A}$ and

$\hat{B}$, that is d(60 ℃)= 8.7164e-4. The estimation value of σ can also get from the estimation value of $\sigma^2 \Delta t$ under 120 °C. Finally, the reliability estimation value of product is obtained by plugging d(60 ℃) and σ in(5), which is shown in Table 5.

Table 5. The Estimation Value of Reliability and Lifetime

	3 year	5 year	7 year	10 year
R(T)	1	1	0.8384	0.0127

5. Conclusion

1) This paper applies Bayesian method to handle degradation data obtained from ADT with a new data processing and parameter inference method.
2) The estimation values and processed data of previous stress level can contribute the estimation under next stress level with updating and iteration to maximize the utilization of test information to gain the best result.
3) The method proposed can eliminate the deviation effect caused by data processing and improve the estimation accuracy and be a useful tool to apply in ADT design optimization.
4) A case study is provided to demonstrate the validity and superiority of the proposed parameter estimation method as well as the reliability estimation method.
5) This paper only considers product which subject single stress, products follow the multiple failure mode would be studied in future.

6. References

[1] Chen S L and Tsai H J. A Novel Adaptive Local Dimming Backlight Control Chip Design Based on Gaussian Distribution for Liquid Crystal Displays[J]. *Journal of Display Technology*, 2017, **12(12)**:1494-1505.

[2] Wang F K and Chu T P. Lifetime predictions of LED-based light bars by accelerated degradation test[J]. *Microelectronics Reliability*, 2012, **52(7)**:1332-36.

[3] Yang F, Wang D, Xing Y and Tsui K L. Prognostics of Li(NiMnCo)O2 -based lithium-ion batteries using a novel battery degradation model[J]. *Microelectronics Reliability*, 2017, **70**:70-78.

[4] Thomas E V, Bloom I, Christophersen J P and Battaglia V S. Statistical methodology for predicting the life of lithium-ion cells via accelerated degradation testing[J]. *Journal of Power Sources*, 2008, **184(1)**:312-7.

[5] Sun F, Liu L, Li X, and Liao H. Stochastic Modeling and Analysis of Multiple Nonlinear Accelerated Degradation Processes through Information Fusion:[J]. *Sensors*, 2016, **16(8)**.

[6] Ye Z and Xie M. Stochastic modelling and analysis of degradation for highly reliable products[J]. *Applied Stochastic Models in Business & Industry*, 2015, **31(1)**:16-32.

[7] Liu L, Li X, Sun F and Wang N. A General Accelerated Degradation Model Based on the Wiener Process.[J]. *Materials*, 2016, **9(12)**:981.

[8] Guan Q, Tang Y and Xu A. Objective Bayesian analysis accelerated degradation test based on Wiener process models ☆[J]. *Applied Mathematical Modelling*, 2016, **40(4)**:2743-55.

[9] Wang H, Xu T and Mi Q. Lifetime prediction based on Gamma processes from accelerated degradation data[J]. *Chinese Journal of Aeronautics*, 2015, **28(1)**:172-9.

[10] Xu A, Basu S and Tang Y. A Full Bayesian Approach for Masked Data in Step-Stress Accelerated Life Testing[J]. *IEEE Transactions on Reliability*, 2014, **63(3)**:798-806.

IOP Conf. Series: Materials Science and Engineering **351** (2018) 012009 doi:10.1088/1757-899X/351/1/012009

Multistage degradation modeling for BLDC motor based on Wiener process

Qingyang Yuan[1], Xiaogang Li[1] and Yuankai Gao[1]

[1]School of reliability and Systems Engineering, Beihang University, BJ, 100191, China

tsyoung.yuan@buaa.edu.cn

Abstract. Brushless DC motors are widely used, and their working temperatures, regarding as degradation processes, are nonlinear and multistage. It is necessary to establish a nonlinear degradation model. In this research, our study was based on accelerated degradation data of motors, which are their working temperatures. A multistage Wiener model was established by using the transition function to modify linear model. The normal weighted average filter (Gauss filter) was used to improve the results of estimation for the model parameters. Then, to maximize likelihood function for parameter estimation, we used numerical optimization method- the simplex method for cycle calculation. Finally, the modeling results show that the degradation mechanism changes during the degradation of the motor with high speed. The effectiveness and rationality of model are verified by comparison of the life distribution with widely used nonlinear Wiener model, as well as a comparison of QQ plots for residual. Finally, predictions for motor life are gained by life distributions in different times calculated by multistage model.

1. Introduction

Brushless DC (BLDC) motors are widely used in a variety of drive and servo systems with good speed control and start characteristics. Wang Lingling[1] used the traditional zero-failure test method to verify the average life of the motors, and gave a test plan to evaluate whether the batch of motors was passed. Due to the long test time, she discussed the method for verifying the average life of an electric motor by accelerated life test. Wang Jian[2] made a useful exploration on the reliability demonstration by accelerated test. The results show that the scheme shortens the test time effectively. Due to the high reliability and long life characteristics of the motor, the life test often can hardly obtain enough failure life data. If the degradation data can be effectively utilized, it can make up the problem of insufficient information on the life expectancy of long life products. The modeling methods based on performance degradation data are effective complement to the traditional life prediction method, which have been successful in theoretical research and engineering application. Among them, Wiener process has a wide range of applications in the degradation modeling. Wang[3] described the degradation process of the product as linear drift Wiener process and proposed an adaptive method for predicting the remaining life of the product. The model was used to describe the degradation process of the bridge. Based on time scale transformation, the degradation processes of transistors and cables were studied by Whitmore[4]. Tseng and Peng established a nonlinear Wiener process model to describe the performance degradation of LED. Si established the nonlinear drift model by Wiener process, using standard Brown motion to describe the uncertainty of degradation on the time axis, then gave the approximate expression of the failure density function and the probability density function of the residual life. Wang Xiaolin[5] summed up the theory of multiple nonlinear degradation modeling, and gave cases for analysis. However, in most cases, as the degradation of the motor continues, its dominant mechanism is changing, resulting in degradation appearing multi-stage, not a single model linear or non-linear with time-scale transformation.

2. Accelerated Degradation Test

Accelerated degradation test technique[6] is method combined of degradation test and accelerated test, by failure mechanism remains unchanged, to extrapolate and predict life under normal stress levels using product performance degradation data at high stress levels. Upon the manufacture’ s request, we carried out an accelerated test and life evaluation for a batch of motors. In this research, the accelerated tests for BLDC motors were loaded at high temperature (75 ℃) and high load stress, and the rise-temperatures were considered the degradation data of the motor.

3. Wiener Process

3.1. Linear wiener process

From the practical point of view, randomness needs to be considered in the degradation modeling: the randomness in the sample performance and the randomness in the degradation on the time axis. The randomness of the samples depicts the samples differences, manifesting different degradation rates; the randomness on the time axis characterizes fluctuations in product degradation (control system fluctuations, product performance fluctuations, observation fluctuations). To this end, the models based on stochastic process are favored by many scholars. If $\{X(t): t \geq 0\}$ satisfies:

① the increments in any time satisfies the normal distribution:

$$DX_i = X(t_{i+1}) - X(t_i),\ Dt_i = t_{i+1} - t_i,\ DX_i \sim N(uDt_i, \sigma^2 Dt_i)$$

Where, u for drift coefficient and σ for diffusion coefficient.

② The increments in any time are independent of each other.

Then the stochastic process[7] is called a linear Wiener process, which is used by Tseng and Tsai to describe the product degradation process. The model is:

$$X(t) = X(0) + ut + \sigma B(t) \tag{1}$$

Where, u for drift coefficient and σ for diffusion coefficient, $B(t)$ for Brownian motion.

We used chi-square test to verify that the temperature data of the accelerated degradation record were normally distributed and chose 50 data before sample starts steady degradation. Results as see Table 1.

Table 1. Data distributions.

temperature range	amount
u+0.3~u+0.4	22
u–0.2~u–0.1, u+0.1~u+0.2	19
u–0.3~u–0.2, u+0.2~u+0.3	8
u–0.4~u–0.3, u+0.3~u+0.4	1
<u–0.4, >u+0.4	0

Its Chi square test is: $\chi^2 = 1.87 < 9.49 = \chi_{0.05}{}^2(4)$. The assumption that degradation data were normally distributed was accepted and the error caused by assumption was acceptable to do the theoretical calculation.

The failure distribution corresponding to this degenerate model is the inverse Gaussian distribution, and the failure probability density function is:

$$f(t) = \frac{D}{\sqrt{2\pi\sigma^2 t^3}} e^{-\frac{(D-ut)^2}{2\sigma^2 t}} \tag{2}$$

Where D is the threshold level of X(t). Coefficients[8,9] are gained by maximum likelihood estimation:

$$\hat{u} = \frac{\sum_{i=1}^{n} \Delta X_i}{\sum_{i=1}^{n} t_i}, \quad \hat{\sigma} = \sqrt{\frac{1}{n}\sum_{i=1}^{n}\frac{(\Delta X_i - \hat{u} t_i)^2}{t_i}}$$

Where, $\Delta X_i = X(t_{i+1}) - X(t_1)$

3.2. *Multistage wiener process*

In this research, a transition function[10] was used:

$$G(\gamma, c; s_t) = \left(1 + \exp(-\gamma(s_t - c))\right)^{-1} \tag{3}$$

Where, s_t is time-related variables, and $s_t = t$ here; c is location parameter of transition; γ is speed parameter of transition.

The multistage Wiener process is expressed as:

$$X(t) = X(0) + (1 - G(\gamma, c; t))u_1 t + G(\gamma, c; t)(u_2 t + x_0) + \sigma B(t) \tag{4}$$

At this point, when degradation process hits the threshold in a certain time, the probability that the process has crossed the threshold level before this time is considered negligible, which is referred by [11]. In fact, the diffusion process - Brown motion may hit threshold many times. With assumption that the probability of this case is 0, the probability density function of the Wiener can be expressed as follows:

$$f(t) \approx \frac{1}{\sqrt{2\pi t}}\left(\frac{S(t)}{t} + \frac{\lambda(t,\theta)}{\sigma}\right)\exp\left(-\frac{S^2(t)}{2t}\right) \tag{5}$$

Where, $S(t) = \frac{1}{\sigma}(D - \lambda(t,\theta))$, $\lambda(t,\theta) = (1 - G(\gamma, c; t))u_1 t + G(\gamma, c; t)(u_2 t + x_0)$, $\theta = (u_1, u_2, \gamma, c, x_0)$ is the parameter vector for λ .

So, its Likelihood function is:

$$L(\theta,\sigma) = \prod_i \frac{1}{\sqrt{2\pi t_i}}\left(\frac{S(DX_i,t_i)}{t_i} + \frac{\lambda(t_i,\theta)}{\sigma}\right)\exp\left(-\frac{S^2(DX_i,t_i)}{2t_i}\right) \tag{6}$$

Where, $S(DX_i, t_i) = \frac{1}{\sigma}(DX_i - \lambda(t_i,\theta))$.

The process of estimating the parameters is the process of solving $\arg\max L(\theta,\sigma)$.

4. Cases Studies

4.1. *Computing method*

We used simplex method[12, 13] to estimate the parameters in the model, which is as follows:

- Establish seven vertices of the regular simplex in the parameter space $\{u_1, u_2, \gamma, c, x_0, \sigma\}$, which is $\{d_1, d_2, d_3, d_4, d_5, d_6, d_7\}$, where $d_i = (u_{1i}, u_{2i}, \gamma_i, c_i, x_{0i}, \sigma_i)$;
- Calculate the logarithmic likelihood of the regular simplex vertices, record the results as $\{l_1, l_2, l_3, l_4, l_5, l_6, l_7\}$, select the smallest l_k , prepare corresponding d_k for deletion;
- Search the backup point : $d_{new} = \frac{1}{n}\sum_{j=1\&j\neq k}^{7} d_j + \left(\frac{1}{n}\sum_{j=1\&j\neq k}^{7} d_j - d_k\right)$;
- Compare the backup point with the worst point of the log likelihood function value $\{l_{new}, l_k\}$, if better then update $d_k = d_{new}$, or shorten the step size, or stop if meet the accuracy required to calculate.

4.2. *Failure probability distribution*

The failure probability distribution is obtained by integrating the failure probability density function[14]:

$$F(t)=\int_t f(t)=\int_t \frac{1}{\sqrt{2\pi t}}\left(\frac{S(t)}{t}+\frac{\lambda(t,\theta)}{\sigma}\right)\exp\left(-\frac{S^2(t)}{2t}\right) \tag{7}$$

We predicted the distribution of life in the future, that is, calculated probability density function at different times:

$$f'(t\,|\,t_0)=\frac{1}{1-F(t_0)}f(t)\,,t\ge t_0 \tag{8}$$

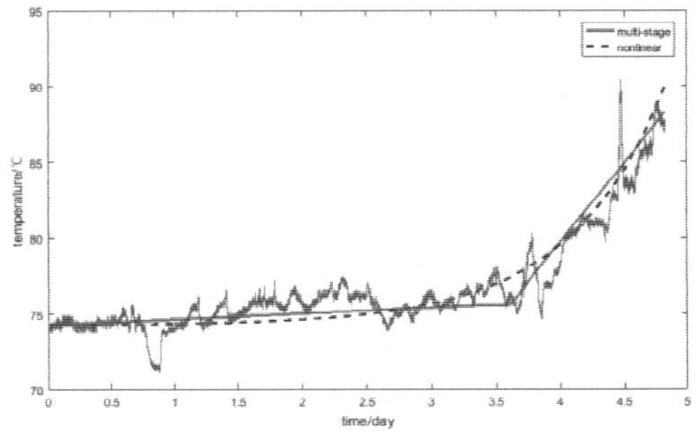

Figure 1. Comparison of Drift Process.

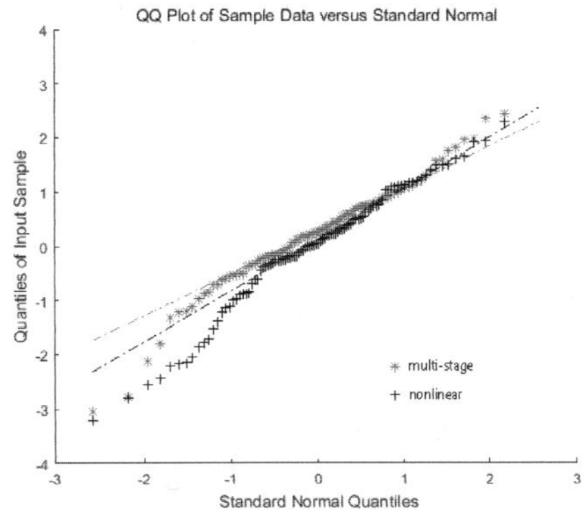

Figure 2. Residual QQ Plot.

4.3. *Results*

Using the method of this paper, we got the drift process as shown in Fig 1, with parameters: $u_1=0.3860$, $u_2=10.3378$, $c=3.6259$, $r=25.9924$, $x_0=-35.8467$, $\sigma=3.7550$. This shows the fact that the mechanism of degradation changes in 3.6 days, and the rate of change is pretty fast. The drift process obtained by the nonlinear method used in [4, 5] is also shown in fig1 as follows. Their Residual QQ Plot[15] are as Figure 2

The comparison of the failure distributions (Figure 3) shows that using nonlinear models to predict the life distribution of the product is earlier than practical. In the final stage, the degradation speed of the nonlinear model is faster than the actual degradation trend, resulting in a poor description for the degradation rate of the later stage of the motor. The failure distribution of the multi-stage degradation process is more optimized, and the description in whole period is also more appropriate. The life prediction of the motor under this stress carried out by using the multistage Wiener process model are shown as Figure 4.

5. Conclusions

In this research, aiming at the degradation characteristics of BLDC motor –nonlinear and multistage, we adapted the Wiener model, which was different from the individual nonlinear process, to research on modeling methods. This method was applied to solve the multi - stage Wiener process modeling problem, which is of great significance to the engineering application for the life prediction of the motor.

- We established the multi-stage Wiener process model to describe BLDC motor degradation. According to the multi-stage model theory proposed in this paper, the mechanism of motor in the process of degradation has changed, and the conversion speed is pretty fast.
- In the process of accelerated life test, attention should be paid to whether the mechanism of degradation changes; when the mechanism changes, the accelerated life test should be carried out with caution.

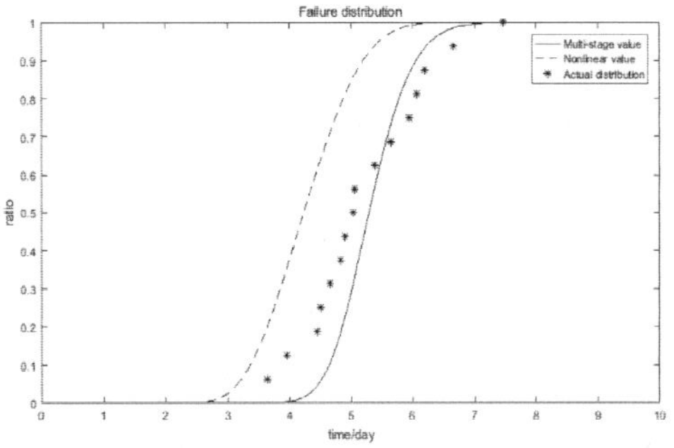

Figure 3. Comparison of Failure Distributions.

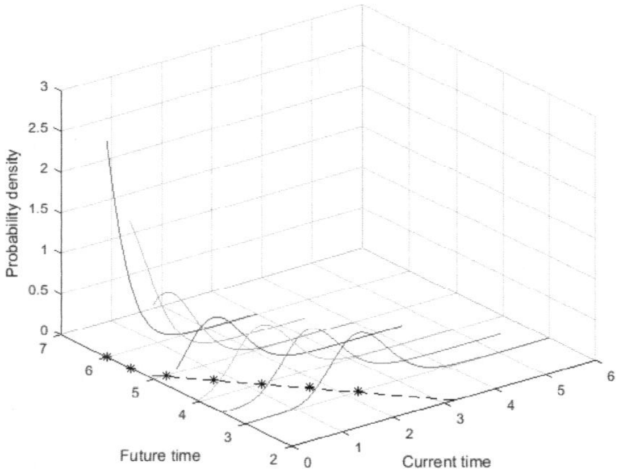

Figure 4. Life Prediction.

6. References

[1] Lingling W, Shisong M and Jing W 2007 Zero failure acceptance scheme for average life of motor *China Academic Journal Electronic Publishing House*.

[2] Jian W The Accelerated Testing to Estimate the Reliability Specifications *Electronic Product Reliability and Environmental Testing* 2010 **28** 6-10

[3] Wang W A model for residual life prediction based on Brownian motion with an adaptive drift *Microeletronics Reliability* 2011 **51** 285-293.

[4] Whitmore G A 1997 Modelling Accelerated Degradation Data Using Wiener Diffusion With A Time Scale Transformation *Lifetime Data Analysis* **26** 623-7

[5] Xiaolin W, Ping J, Yunyan X and Bo G 2015 *Residual Life Estimation for Nonlinear-Deterioriate Products* (Beijing: Defense Industry Press) pp 39-46

[6] Tongmin J 2012 *Reliability Maintainability Supportability* (Beijing: Defense Industry Press)

[7] Lonnie C L 2005 *Random Processes: Filtering, Estimation, and Detection* (Beijing: Publishing House of Electronics Industry)

[8] Chunyan Z 2015 *Research on the Thoery and Application of Smooth Transition Autoregressive Mode* (Beijing: Tsinghua University Press) pp 28-30

[9] Haitao L 2011 Momentum Wheel Wiener Process *Journal of Aerospace Power* **26** 623-7

[10] Sheng S X andWenbinW2012 Remaining Useful Life Estimation Based on a Nonlinear Diffusion Degradation Process *IEEE Transactions on Reliability* **61** 50-67

[11] Tsaing T S and Yu P C 2007 Stochastic Diffusion Modeling of Degradation Data *Journal of Data Science* **5** 315-333

[12] Jorge N, Stephen J W 2006 Wright. *Numerical Optimization* (Beijing: Science Press) pp 363-378

[13] Zheng W 2016 A Improved Quantum Behaved Particle Swarm Optimization Algorithm Using Nelder and Mead's Simplex Algorithm *Complex Sestems and Comples Science* **13** 97-104

[14] Shufeng W 2014 Residual life prediction method of electronic products with Wiener process degradation *Electronic Measurement Technology* **37** 17-20

[15] Zhenhai Y, Weihu C and Junjian Z 2011 *Goodness of test* (Beijing: Science Press)

IOP Conf. Series: Materials Science and Engineering **351** (2018) 012010 doi:10.1088/1757-899X/351/1/012010

Analysis of SET pulses propagation probabilities in sequential circuits

Shuo Cai, Fei Yu and Yiqun Yang

1 School of Computer and Communication Engineering, Changsha University of Science and Technology, Changsha 410114, China
2 Hunan Provincial Key Laboratory of Intelligent Processing of Big Data on Transportation, Changsha University of Science and Technology, Changsha 410114, China

caishuo@csust.edu.cn

Abstract. As the feature size of CMOS transistors scales down, single event transient (SET) has been an important consideration in designing logic circuits. Many researches have been done in analyzing the impact of SET. However, it is difficult to consider numerous factors. We present a new approach for analyzing the SET pulses propagation probabilities (SPPs). It considers all masking effects and uses SET pulses propagation probabilities matrices (SPPMs) to represent the SPPs in current cycle. Based on the matrix union operations, the SPPs in consecutive cycles can be calculated. Experimental results show that our approach is practicable and efficient.

1. Introduction

Due to the reduction in device feature size, transient faults in logic circuits induced by radiations increase dramatically. Transient faults can be categorized into SETs and single event upsets (SEUs) [1, 2]. As technology feature size shrinks, reduced supply voltage and increased frequency result in more SETs being propagated to latches or primary outputs of circuits and creates more failures than before [3, 4].

Sequential circuit elements, namely logic gates (GTs) and flip-flops (FFs), are most susceptible to transient faults [5]. One common approach to analyze the impact of SETs is to inject the pulses at given nodes and simulate the circuit for different input vectors so as to find whether the SET pulses propagate [6]. Nevertheless, it becomes intractable for large circuits and large number of inputs. On the contrast, another type of approaches use the signal probability theory to analyze the SPPs [7]. The computational complexity of these approaches is lower but its accuracy is compromised because the influence of signal correlations have not been considered. Furthermore, above both kinds of approaches usually fail to include at least one of the following: 1) considering all masking effects (logical, electrical and latching-window masking effects). 2) being suitable for sequential circuits working in consecutive cycles. Therefore, better approaches that use concise mathematical tools and reasonable method are required to analyze the SPPs.

In this paper, we present a new approach for analyzing the SPPs in sequential logic circuits. The SPPs of gates and flip-flops in current cycle are represented with SPPMs, then the SPPs in consecutive cycles are calculated by matrix union operation. On the basis of this, the SETs' influence on the soft error rate (SER) and reliability of circuit can be evaluated efficiently.

2. SET pulses propagation probability matrices

SET pulses can be directly propagated to primary outputs (POs) and cause the circuit failure currently, or they can be propagated to flip-flops repeatedly, and finally appear as failure on POs several cycles later. Our definitions refer to the synchronous sequential logic circuit which consists of *m* logic gates and *n* flip-flops.

Definition 1. $\text{SPPM}_{\text{GT-PO}}$ is defined as a $m \times 1$ column vector $[P(\text{PO}|\text{GT}_1)\ P(\text{PO}|\text{GT}_2)\ \ldots\ P(\text{PO}|\text{GT}_m)]^{\text{T}}$, where $P(\text{PO}|\text{GT}_i)(i=1,2,\ldots,m)$ is the average probability of the SET pulse propagated from error site GT_i to POs at current cycle.

We use α, β and γ to represent logical, electrical and latching-window propagation factor of SET pulse respectively. $\alpha_{\text{G}_i\text{P}}^{\text{V}_a}$ $(i=1,2,\ldots,m)$ means the SPP from GT_i to POs avoiding logical masking effect under the stimulus of input vector V_a. Obviously, the value of $\alpha_{\text{G}_i\text{P}}^{\text{V}_a}$ is either 0 or 1. If it is 1, then mark all the sensitive paths from GT_i to reachable POs under the stimulus of V_a and find out the shortest one (labeled as S_{path}). If the SET pulse is electrical masked through S_{path}, then the value of $\beta_{\text{G}_i\text{P}}^{\text{V}_a}$ is 0, otherwise, $\beta_{\text{G}_i\text{P}}^{\text{V}_a}$ is 1. Based on this, $P(\text{PO}|\text{GT}_i)$ can be computed as

$$P(\text{PO}\,|\,\text{GT}_i) = \frac{1}{k}\sum_{a=1}^{k} \alpha_{\text{G}_i\text{P}}^{\text{V}a} \cdot \beta_{\text{G}_i\text{P}}^{\text{V}a} \tag{1}$$

where k is the number of applied input vectors.

Definition 2. $\text{SPPM}_{\text{GT-FF}}$ is defined as a $m \times n$ matrix $\begin{bmatrix} P(\text{FF}_1\,|\,\text{GT}_1) & \ldots & P(\text{FF}_n\,|\,\text{GT}_1) \\ \ldots & \ldots & \ldots \\ P(\text{FF}_1\,|\,\text{GT}_m) & \ldots & P(\text{FF}_n\,|\,\text{GT}_m) \end{bmatrix}$, where $P(\text{FF}_j|\text{GT}_i)(i=1,2,\ldots,m;\ j=1,2,\ldots,n)$ is the average probability of the SET pulse propagated from error site GT_i to FF_j and latched by FF_j at current cycle.

We use $\alpha_{\text{G}_i\text{F}_j}^{\text{V}_a}$, $\beta_{\text{G}_i\text{F}_j}^{\text{V}_a}$ and $\gamma_{\text{G}_i\text{F}_j}^{\text{V}_a}$ to represent the SPP from GT_i to FF_j avoiding logical, electrical and latching-window masking effects under the stimulus of input vector V_a respectively. If $\alpha_{\text{G}_i\text{F}_j}^{\text{V}_a}$ and $\beta_{\text{G}_i\text{F}_j}^{\text{V}_a}$ are both 1, then update the width of SET pulse and calculate $\gamma_{\text{G}_i\text{F}_j}^{\text{V}_a}$ further. The matrix element $P(\text{FF}_j|\text{GT}_i)$ can be expressed as

$$P(\text{FF}_j\,|\,\text{GT}_i) = \frac{1}{k}\sum_{a=1}^{k} \alpha_{\text{G}_i\text{F}_j}^{\text{V}a} \cdot \beta_{\text{G}_i\text{F}_j}^{\text{V}a} \cdot \gamma_{\text{G}_i\text{F}_j}^{\text{V}a} \tag{2}$$

Definition 3. $\text{SPPM}_{\text{FF-FF}}$ is defined as a $n \times n$ matrix $\begin{bmatrix} P(\text{FF}_1\,|\,\text{FF}_1) & \ldots & P(\text{FF}_n\,|\,\text{FF}_1) \\ \ldots & \ldots & \ldots \\ P(\text{FF}_1\,|\,\text{FF}_n) & \ldots & P(\text{FF}_n\,|\,\text{FF}_n) \end{bmatrix}$, where $P(\text{FF}_j|\text{FF}_i)(i, j=1,2,\ldots,n)$ is the average probability of the SET pulse propagated from error site FF_i to FF_j and latched by FF_j at current cycle. By considering three masking effects, $P(\text{FF}_j|\text{FF}_i)$ can be expressed as

$$P(\text{FF}_j\,|\,\text{FF}_i) = \frac{1}{k}\sum_{a=1}^{k} \alpha_{\text{F}_i\text{F}_j}^{\text{V}a} \cdot \beta_{\text{F}_i\text{F}_j}^{\text{V}a} \cdot \gamma_{\text{F}_i\text{F}_j}^{\text{V}a} \tag{3}$$

We also define $\text{SPPM}^*_{\text{FF-FF}}$ as $\begin{bmatrix} P^*(\text{FF}_1\,|\,\text{FF}_1) & \ldots & P^*(\text{FF}_n\,|\,\text{FF}_1) \\ \ldots & \ldots & \ldots \\ P^*(\text{FF}_1\,|\,\text{FF}_n) & \ldots & P^*(\text{FF}_n\,|\,\text{FF}_n) \end{bmatrix}$, where $P^*(\text{FF}_j|\text{FF}_i)$ is the probability of the pulse latched by FF_i propagated from FF_i to FF_j and latched by FF_j at next cycle. Here, β and γ are different from Eq.(3) because of the variation of initial SET pulse width. More specifically, the pulse width has been changed to a complete clock cycle. The expression can be written as

$$P^*(\text{FF}_j\,|\,\text{FF}_i) = \frac{1}{k}\sum_{a=1}^{k} \alpha_{\text{F}_i\text{F}_j}^{\text{V}a} \cdot \beta_{\text{F}_i\text{F}_j}^{*\text{V}a} \cdot \gamma_{\text{F}_i\text{F}_j}^{*\text{V}a} \tag{4}$$

Definition 4. $\text{SPPM}_{\text{FF-PO}}$ is defined as a $n \times 1$ column vector $[P(\text{PO}|\text{FF}_1)\ P(\text{PO}|\text{FF}_2)\ \ldots\ P(\text{PO}|\text{FF}_n)]^{\text{T}}$, where $P(\text{PO}|\text{FF}_i)$ $(i=1,2,\ldots,n)$ is the average probability of the SET pulse propagated from FF_i to POs at current cycle. It can be achieved as

$$P(\mathrm{PO}\,|\,\mathrm{FF}_i)=\frac{1}{k}\sum_{a=1}^{k}\alpha_{\mathrm{F}_i\mathrm{P}}^{\mathrm{V}a}\cdot\beta_{\mathrm{F}_i\mathrm{P}}^{\mathrm{V}a} \tag{5}$$

We define $\mathrm{SPPM}^*_{\text{FF-PO}}$ as $[P^*(\mathrm{PO}|\mathrm{FF}_1)\ P^*(\mathrm{PO}|\mathrm{FF}_2)\ ...\ P^*(\mathrm{PO}|\mathrm{FF}_n)]^{\mathrm{T}}$, where $P^*(\mathrm{PO}|\mathrm{FF}_i)$ is the probability of the pulse latched by FF_i propagated from FF_i to POs at next cycle, and $P^*(\mathrm{PO}|\mathrm{FF}_i)$ can be written as

$$P^*(\mathrm{PO}\,|\,\mathrm{FF}_i)=\frac{1}{k}\sum_{a=1}^{k}\alpha_{\mathrm{F}_i\mathrm{P}}^{\mathrm{V}a}\cdot\beta_{\mathrm{F}_i\mathrm{P}}^{*\mathrm{V}a} \tag{6}$$

The above SPPMs can be obtained in advance. To obtain α in Eq.(1-6), a simulation-based method is adopted. This method injects SET pulse per each node respectively and applies certain vector randomly. Afterwards, α can be computed by simulating the circuit and comparing the logical values to the correct ones of POs or specific FF. For a specific SET pulse, if its corresponding $\alpha^{\mathrm{V}_a}=1(a=1,2,...,k)$, then find out the shortest path from all sensitive paths between fault site to reachable PO or FF under the stimulus of V_a (If the pulse can be electrical masked through the shortest path, it can also be electrical masked through any other sensitive paths). For a gate delay of D_{GT} and pulse width of W_{pi} at the gate input, pulse width at the output of the gate, W_{po}, can be approximated as[8]

$$W_{\mathrm{po}}=\begin{cases}0, & W_{\mathrm{pi}}<D_{\mathrm{GT}}\\ 2(W_{\mathrm{pi}}-D_{\mathrm{GT}}), & D_{\mathrm{GT}}\le W_{\mathrm{pi}}<2D_{\mathrm{GT}}\\ W_{\mathrm{pi}}, & W_{\mathrm{pi}}\ge 2D_{\mathrm{GT}}\end{cases} \tag{7}$$

For a given SET pulse, if α^{V_a} and β^{V_a} are both equal to 1, γ can be evaluated by Eq.(8). Here, T_{c} is the clock period, T_{setup} and T_{hold} are the setup time and hold time of the latching flip-flop respectively [9]. Given W_{pi}, D_{GT}, T_{setup}, T_{hold} and T_{c}, the elements of SPPMs can be obtained by applying k test vectors and using Eq. (1-8).

$$\gamma=\begin{cases}0, & W_{\mathrm{po}}<T_{\mathrm{setup}}+T_{\mathrm{hold}}\\ \dfrac{W_{\mathrm{po}}-(T_{\mathrm{setup}}+T_{\mathrm{hold}})}{T_{\mathrm{c}}}, & W_{\mathrm{po}}\ge T_{\mathrm{setup}}+T_{\mathrm{hold}}\end{cases} \tag{8}$$

Besides that, based on Eq. (7-8), we find that β^* and γ^* are approximated 1 when the clock frequency is lower than 20GHz. Therefore, Eq. (4) and Eq. (6) can be simplified into $P^*(\mathrm{FF}_j\,|\,\mathrm{FF}_i)=\frac{1}{k}\sum_{a=1}^{k}\alpha_{\mathrm{F}_i\mathrm{F}_j}^{\mathrm{V}a}$ and $P^*(\mathrm{PO}\,|\,\mathrm{FF}_i)=\frac{1}{k}\sum_{a=1}^{k}\alpha_{\mathrm{F}_i\mathrm{P}}^{\mathrm{V}a}$ respectively.

3. SPPs analysis in consecutive clock cycles

Definition 5. The operator of matrix union operation is expressed as "$\cup$". If the number of columns of matrix M is equal to the number of rows of matrix N, then the matrix union operation can be described as follows:

$$\mathrm{M}_{r\times s}=\begin{bmatrix}m_{11} & \cdots & m_{1s}\\ \cdots & m_{ij} & \cdots\\ m_{r1} & \cdots & m_{rs}\end{bmatrix},\ \mathrm{N}_{s\times t}=\begin{bmatrix}n_{11} & \cdots & n_{1t}\\ \cdots & n_{ij} & \cdots\\ n_{s1} & \cdots & n_{st}\end{bmatrix},\ \mathrm{M}_{r\times s}\cup\mathrm{N}_{s\times t}=\mathrm{L}_{r\times t}=\begin{bmatrix}l_{11} & \cdots & l_{1t}\\ \cdots & l_{ij} & \cdots\\ l_{r1} & \cdots & l_{rt}\end{bmatrix},$$

where $0\le m_{ij}, n_{ij}\le 1(m_{ij}\in\mathrm{M}, n_{ij}\in\mathrm{N})$ and $l_{ij}=1-\prod_{k=1}^{s}(1-m_{ik}n_{kj}),(1\le i\le r,1\le j\le t)$

Definition 6. The matrix exponentiation operation based on union operation is described as follows: $\mathrm{M}^{\bar{c}}=\mathrm{M}\cup\mathrm{M}\cup...\cup\mathrm{M}$ (the number of M is c).

Theorem1. The probability of a circuit failure at cth cycle given a SET occurs at GT_i (the time for pulse occurring is 1st cycle), $\mathrm{P}^c(\mathrm{PO}|\mathrm{GT}_i)$, is described as:

$$P^c(\mathrm{PO}\,|\,\mathrm{GT}_i)=\begin{cases}\mathrm{Row}_i(\mathrm{SPPM}_{\text{GT-PO}}), & c=1\\ \mathrm{Row}_i(\mathrm{SPPM}_{\text{GT-FF}}\cup(\mathrm{SPPM}^*_{\text{FF-FF}})^{\overline{c-2}}\cup\mathrm{SPPM}^*_{\text{FF-PO}}), & c\ge 2\end{cases} \tag{9}$$

Proof. Here, we consider that the fault propagation to FFs are approximately independent. If $c=1$, $P^1(\mathrm{PO}|\mathrm{GT}_i)=i^{\mathrm{th}}$ row of $\mathrm{SPPM}_{\text{GT-PO}}$; If $c=2$, assume that a SET pulse occurs at GT_i at 1st cycle, which can be propagated to any flip-flop(e.g. FF_j) at current cycle, and then propagated from FF_j to POs at next cycle. That is: $P^2(\mathrm{PO}|\mathrm{GT}_i) = 1-\prod_{j=1}^{n}[1-P(\mathrm{FF}_j\,|\,\mathrm{GT}_i)P^*(\mathrm{PO}\,|\,\mathrm{FF}_j)] = i^{\mathrm{th}}$ row of $(\mathrm{SPPM}_{\mathrm{GT\text{-}FF}} \cup \mathrm{SPPM}^*_{\mathrm{FF\text{-}PO}})$.

The SET pulse can also be propagated to FF_j currently, then propagated to FF_k after one cycle, and to POs after two cycles. Assume $P^2(\mathrm{FF}_j|\mathrm{GT}_i)$ is the probability that the SET pulse is propagated to FF_j one cycle after the pulse occurs given that GT_i is erroneous, and it is the ith row, j^{th} column element of $(\mathrm{SPPM}_{\mathrm{GT\text{-}FF}} \cup \mathrm{SPPM}^*_{\mathrm{FF\text{-}FF}})$. Additionally, $P^3(\mathrm{PO}|\mathrm{GT}_i) = i^{\mathrm{th}}$ row of $(\mathrm{SPPM}_{\mathrm{GT\text{-}FF}} \cup \overline{(\mathrm{SPPM}^*_{\mathrm{FF\text{-}FF}})^{1}} \cup \mathrm{SPPM}^*_{\mathrm{FF\text{-}PO}})$. Similarly, $P^3(\mathrm{FF}_j|\mathrm{GT}_i)$ is the probability that the SET pulse is propagated to FF_j two cycles after the pulse occurs given that GT_i is erroneous. Based on the induction, $P^c(\mathrm{PO}|\mathrm{GT}_i)$ is i^{th} row of $(\mathrm{SPPM}_{\mathrm{GT\text{-}FF}} \cup \overline{(\mathrm{SPPM}^*_{\mathrm{FF\text{-}FF}})^{c-2}} \cup \mathrm{SPPM}^*_{\mathrm{FF\text{-}PO}})$.

Theorem2. The probability of a circuit failure at cth cycle given a SET occurs at FF_i(the time for pulse occurring is 1st cycle), $P^c(\mathrm{PO}|\mathrm{FF}_i)$, is described as:

$$P^c(\mathrm{PO}\,|\,\mathrm{FF}_i) = \begin{cases} \mathrm{Row}_i(\mathrm{SPPM}_{\mathrm{FF\text{-}PO}}), & c=1 \\ \mathrm{Row}_i(\mathrm{SPPM}_{\mathrm{FF\text{-}FF}} \cup \overline{(\mathrm{SPPM}^*_{\mathrm{FF\text{-}FF}})^{c-2}} \cup \mathrm{SPPM}^*_{\mathrm{FF\text{-}PO}}), & c\geq 2 \end{cases} \quad (10)$$

The proving process is similar to that of Theorem 1.

4. Experimental results

Experimental results in the ISCAS'89 circuits are given in this section. We use NANGATE 45-nm technology models to compute the SPPs [10]. V_{DD} is 1.0V, T_{c} is 1ns, D_{GT} is 20ps, $T_{\mathrm{sh}}(T_{\mathrm{setup}}+T_{\mathrm{hold}})$ is 30ps. All experiments have been performed on the system equipped with Intel i3-3110M processor and 4GB memory.

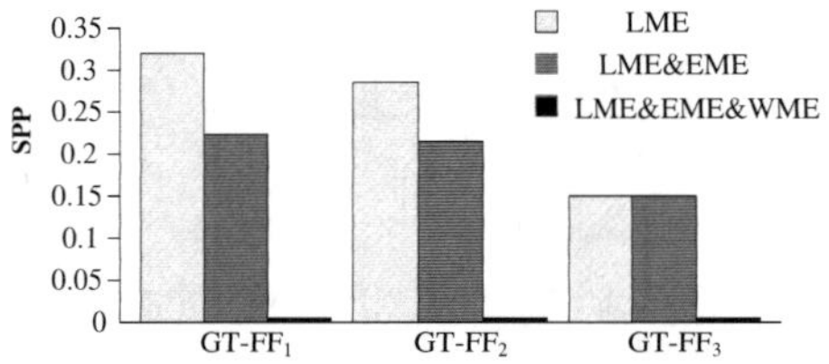

Figure 1. SPPs (GT-FFs at current cycle) with different masking effect considering

Figure 1 shows the average SPPs from GT to FFs of S27 when W_{pi} is 35ps at current cycle. LME, EME and WME represent the logical, electrical and latching-window masking effect respectively. If W_{pi} changes, the SPPs will change accordingly. Figure 2 shows the average SPPs from GT to FFs of S27 at current cycle with different W_{pi} when considering all masking effects.

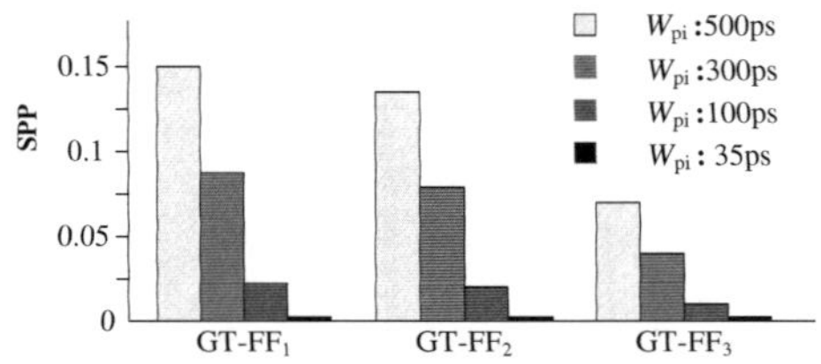

Figure 2. SPPs (GT-FFs at current cycle) with different W_{pi} considering all masking effects

Based on our theorems, we have found the average SPPs from GT (FF) to POs will convergent with clock cycles extend. Figure 3 shows the average SPPs and corresponding calculation times of some ISCAS'89 circuits in consecutive cycles when W_{pi} is 100ps. Figure 4 shows the number of applied vectors to circuits in our experiments and their convergence cycles. Note that the right Y-axis in Figure 3 and left Y-axis in Figure 4 are logarithmic.

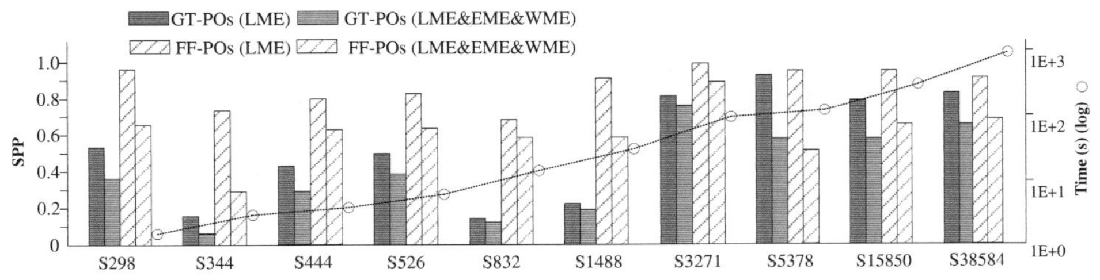

Figure 3. Average SPPs and runtimes of ISCAS'89 in consecutive cycles

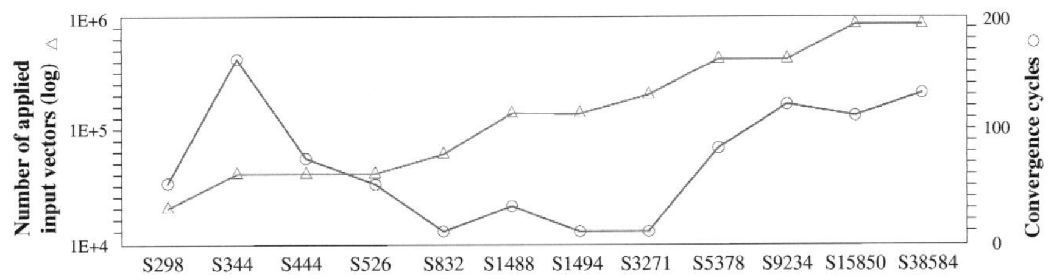

Figure 4. Number of excitation vectors and convergence cycles for experimental circuits

The accuracy of our approach is more than 99.4% compared with Monte Carlo simulation techniques, while its run time is 2 to 3 order magnitudes shorter than that of the latter for experimental circuits[6]. For a sequential circuit with amounts of m gates and n flip-flops, assume the number of applied input vectors is k. In Monte Carlo simulation, the total number of input vectors is k^c when considering the SET pulse propagation during c clock cycles [7, 11]. For a specified SET pulse, the simulation feeds all k^c input vectors and traverses $(m+n)$ elements for each vector to get the result. Therefore, the computational complexity of Monte Carlo simulation is $O(k^c \times (m+n))$. For proposed approach, the SPPs in sequential circuit is obtained by unioning an $n \times n$ matrix c times. So the computational complexity of our approach is $O(c \times n^3)$. It can be seen that the acceleration of our approach will increase with the continuation of clock cycles for larger circuits.

5. Conclusions

In this paper, we have proposed a new SPPs analysis approach in sequential circuits. The approach has two main features: 1) all masking effects that affect the SPPs are considered; 2) the SPPs can be calculated in consecutive cycles efficiently based on SPPMs. These two features make our approach applicable to the reliability evaluation of circuits.

6. Acknowledgment

This paper was supported in part by the National Natural Science Foundation of China (NSFC) under grant No. 61702052, 61504013, in part by the Open Research Fund of Hunan Provincial Key Laboratory of Intelligent Processing of Big Data on Transportation and in part by Hunan Province Undergraduates Innovating Experimentation Project of China under grant No. (2016)283-192.

7. References

[1] N. George, J. Lach. Characterization of logic masking and error propagation in combinational circuits and effects on system vulnerability. *Proc. IEEE/IFIP on DSN*, 2011, 323-330.

[2] M. Grosso. Exploiting fault model correlations to accelerate SEU sensitivity assessment. *IEEE Transactions on Industrial Informatics*, 2013, 1(9):142-148.

[3] B. Narasimham, B. Bhuva, R. Schrimpf, et al. Characterization of digital signal event transient pulse-widths in 130-nm and 90-nm CMOS technologies. *IEEE Transactions on Nuclear Science*, 2007, 6(54):2506-2511.

[4] Hoda P. Systematic analyses for latching probability of single-event transients. *15th Int'l Symposium on Quality Electronic Design*, 2014, 442-449.

[5] K. Mohammadi. Fast reliability analysis method for sequential logic circuits. *21st Int'l Conference on Systems Engineering*, 2011, 352-356.

[6] J. Cong. LUT-based FPGA technology mapping for reliability. *In Proc. IEEE/ACM Design Autom*, 2010, 517-522.

[7] Keheng Huang, Yu Hu, Xiaowei Li. Reliability-oriented placement and routing algorithm for SRAM-based FPGAs. *IEEE Transactions on Very Large Scale Integration (VLSI) Systems*, 2014, 2(22):256-269.

[8] Dhillon Y S, Diril A U, Chatterjee A, Singh A D. Analysis and optimization of nanometer CMOS circuits for soft-error tolerance. *IEEE Transactions on Very Large Scale Integration (VLSI) Systems*, 2006, 5(14):514-524.

[9] Rajeev R. Rao, Kaviraj Chopra, David T. Blaauw, Dennis M. Sylvester. Computing the soft error rate of a combinational logic circuit using parameterized descriptors. *IEEE Transactions on Computer-Aided Design of Integrated Circuits and Systems*, 2007, 3(26):468-479.

[10] Salman E. Reducing delay uncertainty in deeply scaled integrated circuits using interdependent timing constraints. *Int'l workshop on timing issues in the specification and synthesis of digital systems*, 2010, 1-6.

[11] Y. Lin, L. He. Device and architecture concurrent optimization for FPGA transient soft error rate. *Proc. IEEE/ACM Int. Conf. Comput.-Aided Design*, 2007, 194-198.

IOP Conf. Series: Materials Science and Engineering **351** (2018) 012011 doi:10.1088/1757-899X/351/1/012011

Survey of key technologies on millimeter-wave CMOS integrated circuits

Fei Yu[1], Lei Gao[1], Lixiang Li[1], Shuo Cai[1], Wei Wang[1] and Chunhua Wang[2]

[1] Hunan Provincial Key Laboratory of Intelligent Processing of Big Data on Transportation & School of Computer and Communication Engineering, Changsha University of Science and Technology, Changsha 410114, China
[2] College of Information Science and Engineering, Hunan University, Changsha, 410082, China

yufeiyfyf@163.com

Abstract. In order to provide guidance for the development of high performance millimeter-wave complementary metal oxide semiconductor (MMW-CMOS) integrated circuits (IC), this paper provides a survey of key technologies on MMW-CMOS IC. Technical background of MMW wireless communications is described. Then the recent development of the critical technologies of the MMW-CMOS IC are introduced in detail and compared. A summarization is given, and the development prospects on MMW-CMOS IC are also discussed.

1. Introduction

Millimeter-wave (MMW) is an electromagnetic wave between the microwave and the far-infrared wave, the millimeter-wave band is 30-300 GHz. MMW has great significance in communication, radar, navigation, remote sensing, radio astronomy, and biomedical spectroscopy. With the rapid development of microelectronics technology and the growth of military demand, millimeter-wave monolithic integrated circuit (MMIC) became a new type of integrated circuit that integrates multiple field-effect transistors and other components on a single chip.

In order to meet the requirements of high-speed wireless applications, such as high-speed wireless personal area network, vehicular radar, Gigabit point to point link connection [1] etc. The 60 GHz band have been opened despite without international non-license, such as 59-66 GHz in Japan and 57-64 GHz in the United States. For III-V semiconductor technology, such as GaAs and InP, although it has high electron mobility, resistivity and superior performance, the silica-based complementary metal oxide semiconductor (CMOS) has mature progress with low power consumption and low cost. The maximum oscillation frequency fmax and the characteristic frequency ft has been improved with the CMOS process size decreasing, which make the application of CMOS process of the millimeter wave turn possible [2] According to the International Semiconductor Technology Blueprint (ITRS) that give the RF/Analog CMOS evolution route, the characteristic frequency of the CMOS process exceeded the InP in 2009, exceeding the SiGe in 2014 [3], as shown in Figure 1. With the CMOS technology characteristics of the size of the shrinking and working frequency improved, making the millimeter-wave CMOS integrated circuit technology continues to upgrade, it has become a hot spot in the field of MMIC.

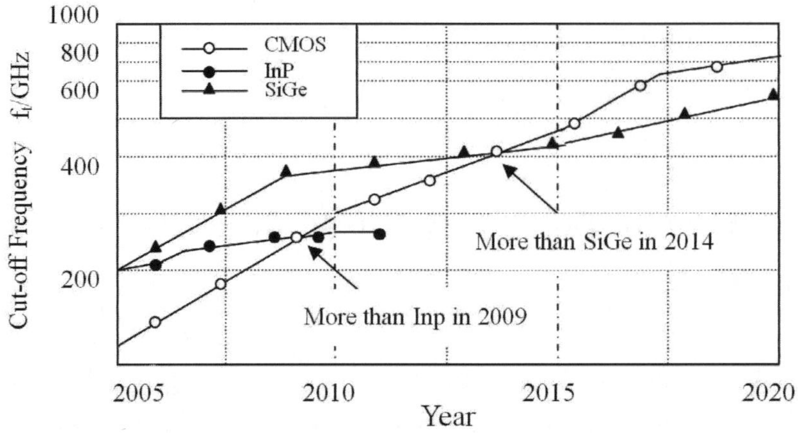

Figure 1. Frequency characteristics of CMOS devices.

The key technology to research the design of MMW-CMOS integrated circuits is to establish accurate and effective CMOS active and passive device models. The ordinary device model will inaccurate when the working frequency of the CMOS process enters the millimeter wave band. The modeling of MMW-CMOS IC mainly includes CMOS field-effect transistor modeling and transmission line modeling which include microstrip line and coplanar waveguide (CPW) modeling etc. This paper summarized key technologies of MMW-CMOS IC, introducing wireless communication technology in the field of millimeter wave, and describing several key techniques for MMW-CMOS IC, and then, the key technologies of MMW-CMOS IC are introduced and compared in detail, furthermore, the future researching direction of the MMW-CMOS IC is prospected and the goal of the MMW-CMOS IC is reviewed .

2. Research on key technologies of MMW-CMOS IC

2.1. Research on de-embedding method in microwave measurement

In the microwave circuit, it is necessary for us to accurate measurement of related devices. In order to get accurate measurement results, it is important that design accurate fixture, meanwhile, the parameters of the device must be deducted from the fixture parameters, which can be called to de-embedding. Using the equivalent circuit to characterize the fixture parameters, this embedding method is point at production of millimeter-wave integrated circuit that base on Inpand GaAs process. With the operating frequency of the integrated circuit reaching the microwave frequency band, a variety of parasitic parameters and skin effect become obvious, thus, it is difficult to use a simple equivalent circuit structure to accurately characterize fixture parameters [4], especially for CMOS integrated circuits, the traditional de-embedding method cannot accurately reflect the microwave characteristics of silicon-based devices, we need to improve the method in the relevant device modeling.

The method of de-embedding including Line-Reflect-Line (LRL) [4], Short-Open-Load-Reciprocal (SOLR) [5], Thru-Reflect-Line (TRL) [6], Line-Reflect-Match (LRM) [7], with the rapid development of CMOS technology, the traditional method of de-embedding is no longer sutable. In 2000, Kolding [8] introduced four-step CMOS device de-embedding, it takes into account substrate loss and contact effects. Using simple texting structure to complete four steps containing open circuit, simple short circuit, open circuit,and short circuit, we can get accurate result. Rosales *et al.* [9] proposed a de-embedding method based on on-chip spiral inductor modeling, the method employ open, left-short, right-short and penetrated four virtual structures to achieve function, and finally using 0.25 μm CMOS process to verify its accuracy. Niu *et al.* [10] compared the four-port de-embedding and verifying it in the frequency range of 2-110 GHz with a CMOS process, and using the NMOS as a DUT to test, as shown in Figure 2. Issakov *et al.* [11] used a new two-port de-embedding technique for differential device modeling and fabricate 2:1 differential transformers and on-chip Thru-Line (TL) with 0.13 μm CMOS process, Thru-Lne-Reflect (TLR) virtual structure, and then used the de-embedding of TL and

TLR to test the transformer S parameters. South Korea's Samsung Electronics related researchers [12] proposed a five-step de-embedding (Pad, Pad-Short, Pad-Open, Short, Open), the five virtual structures were verified by gate electrode resistance and input capacitance of transistor based on 45 nm CMOS process, the results show that the five optimized virtual structures can eliminate the parasitic parameters which generated by pad and metal traces.

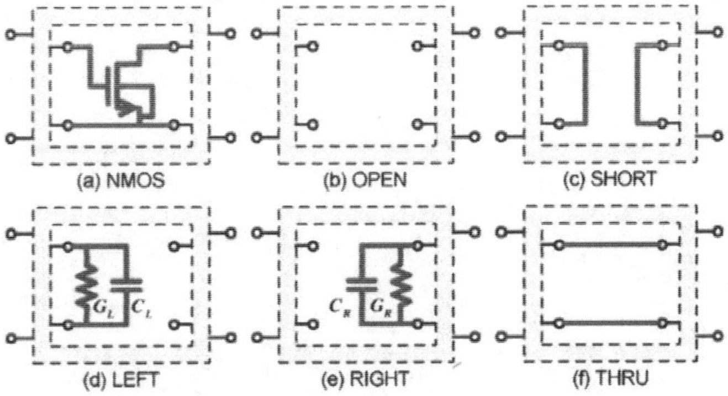

Figure 2. The proposed four-port de-embedding method in Ref. [9].

2.2. Research on modeling method of MOS field effect

The traditional CMOS model provides accurate simulation results in low frequency bands, such as BSIM3 (developed by the University of California, Berkeley Device Research Group), and MOS9 [1] (developed at the Philips Research Laboratory in the Netherlands). However, these CMOS transistor model simulation results appear to large error when the frequency reach to the millimeter-wave band, so how to apply them to the millimeter-wave band what is a very big challenge [2].

It has been theoretically analyzed that f_{max} is independent with the fingers number of multi-finger transistor, so the maximum operating frequency f_{max} of the transistor can be optimized with a single-finger transistor, the layout as shown in Figure 3. A series of resistance and inductive parasitic parameters will become more important when the operating frequency in the microwave frequency band, therefore, these parasitic parameters should be considered in the transistor modeling, the impact of capacitance also cannot be ignored. In the microwave frequency band, the transistor model working is generally divided into the external parameter part and the part of kernel, the external parameter part is the parasitic parameter generated at the high frequency, the part of kernel is the traditional transistor model, as shown in Figure 4. After the model is established, the parameters of the CMOS transistor are extracted by Agilent IC-CAP [13] and Agilent ADS Momentum electromagnetic simulation [14]. The accuracy of the former is higher than the latter owing to necessarily testing transistor layout. The latter parameters need to be extracted directly from the measured data, but the bias conditions are limited.

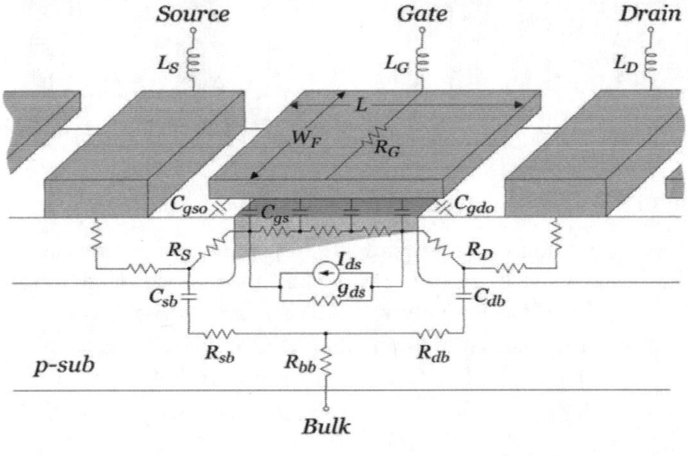

Figure 3. The single-finger physical model of NMOS devices.

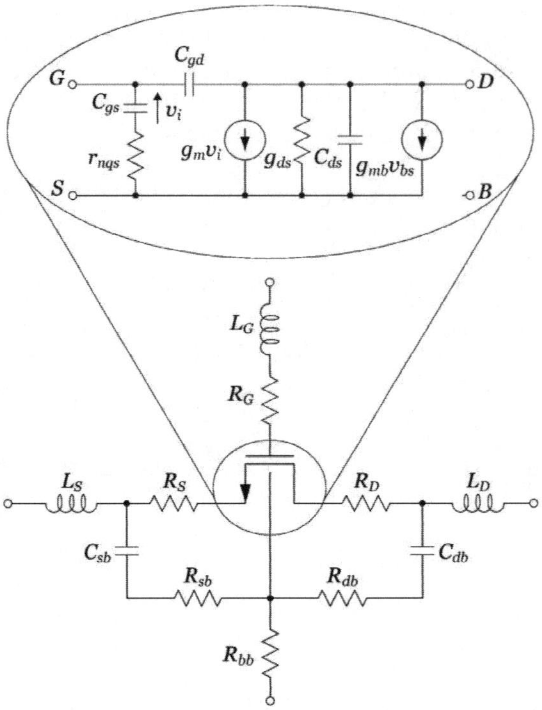

Figure 4. The small signal equivalent model of NMOS device.

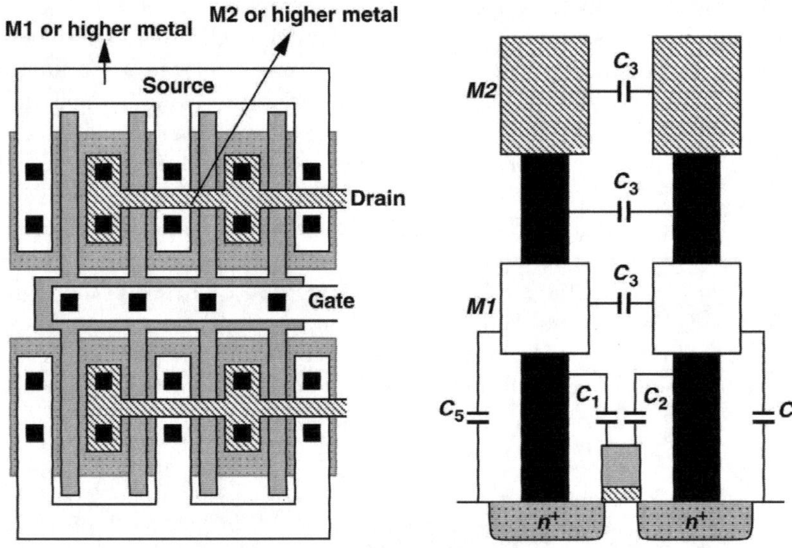

(a) (b)

Figure 5. (a) The Layout of offolding MOS, (b) Section view.

At present, people have made a relatively rich research on the CMOS transistor modeling, as early as 2001, Hsu *et al.* [15] optimized the layout of CMOS and introduced deep wells, which make and of 0.18 μm CMOS process transistor respectively reach to 70 GHz and 58 GHz. In 2007, Zhu *et al.* [16] directly extracted the values of the parasitic elements from the physical map of the MOSFET, and the layout was designed by using the NMOS field-effect transistor of 0.18 μ m CMOS process. The simulation shows that the model can achieve high-frequency performanceof NMOS from 100 MHz to 40 GHz, Dr. Liang and Professor Razavi [17] proposed a folding transistor of physical model, as shown in Figure 5 (a), the model describes the metal capacitance between the transistor stages, as shown in Figure 5 (b), through the new layout technology to reduce these capacitors which can improve the performance of the circuit.

2.3. Research on modeling transmission line

In the millimeter-wave CMOS integrated circuit, the main structure of the transmission line is the microstripline (MSL) and Coplanar Waveguide (CPW). Since the traditional transmission lines are primarily modeled on processes such as GaAs, the principle of propagation will be different from the CMOS process. Therefore, it is particularly important to model the standard CMOS process (the substrate is silicon) transmission line, which will be described in detail. After model is completed, it is necessary to establish the equivalent model of RLGC parasitic parameters, and finally use the electromagnetic simulation software (ADS or HFSS) to extract the parameters, the characteristic impedance range, quality factor Q and other parameters.

2.3.1. Research on modeling method of MSL.

MSL is a planar transmission line. Compared with CPW, it has wider band, higher reliability, smaller size, light weight and lower manufacturing costs, however, the capacity of power is smaller and larger power consumption. Si-based microstrip line use the underlying metal as the ground plane and the top metal as a signal line, substrate has a SiO_2 passivation layer protection. The MSL was manufactured by standard CMOS process which have biggest drawback the signal line is close with the groundsurface, it makes the distributed inductance turn small and reducing the inductance Q.

When the Si-based CMOS process achieve active and passive devices, due to the high loss of silicon substrate between the interaction of electromagnetic field, resulting a large number of substrate losses, to improve it, the relatively thin electrolyte passivation layer was used to modify the traditional transmission line (usually using SiO_2), and other also can be used such as benzo cyclobutene (BCB) and polyimide etc. Wu *et al.* [18] improved the traditional thin film transmission line by placing the Si substrate under the ground, and put four pads on the dielectric layer (passivation layer) for measuring. In order to research the lumped parameters of the relevant TFML, they use less than 1V voltage to measuring and get accurate parameter values. Lee *et al.* [19] proposed anexternal processing method based on Magnetorheological Fluid, which can achieve lower loss of high frequency transmission lines. This paper proposed two transmission lines with 20 μm polyimide dielectric layer by magnetorheological surface processing method, which can effectively reduce the loss but raising processing expense. Chiang *et al.* [20] designed a Quasi-TEM transmission line by using a 0.18 μm 1P6M CMOS process, which was called a Complementary-conducting-strip transmission line (CCS TL). The CCS TL unit consistedby six layers of metal and one layer of dielectric (MD), and then used a number of CCS TL units to form a comprehensive zigzag CCS TL. The simulation showed that the transmission line can obtain wide impedance range (8.62-104.0 Ω), the maximum value of slow wave factor was 4.79, the transmission line area and the quality factor Q ratio was constant when the transmission line length was determined. In 2015, Sang *et al.* [21] proposed a compass transmission line, which characterized with a multi-metallization 65 nm CMOS process, the model, achieving 50 Ω characteristic impedance, getting coefficient of about -27 dB up to 80 GHz and reducing lower insertion loss. In 2017, Zhu [22] introduced anarrowed dielectric microstrip line which put concentrate a dielectric transmission line that consisted by three dielectric substrate layers. Its excellent performance such as ease of fabrication, and low loss make it become suitable for MMW applications.

2.3.2. Research on modeling method of CPW

CPW, all conductors are connected in the same plane, easily realizing the connect with active and passive device. Its propagation mode belongs to quasi-horizontal electromagnetic mode (Q-TEM), CPW has no cut-off frequency, the impedance characteristics is independence of dielectric layer, which is determined by the ratio of the signal line width to the seam widthso that you can freely design its size, compared with microstrip line, it has many advantages such as the smaller dispersion, less loss, higher efficiency. Based on the optimized slot-type floating shields technology of which slow-wave CPW structure is proposed by TSMC and related researchers in Taiwan Jiaotong University with 45 nm CMOS process [23], as shown in Figure 6 (a). The first study shows that the propagation wavelength, attenuation loss and characteristic impedance of CPW can be adjusted by changing the protective layer length (SL), band spacing (SS) and protective layer metal type. The measurement results show that the CPW with the gap-type floating protection structure can reduce the silicon-based area by 67%, compared with the traditional structure, there are more than 9 effective dielectric constants and the quality factor more than 6. In order to obtain a higher quality factor, Kaddour *et al.* [24] simplified the slow-wave CPW structure with gap-type floating protection, as shown in Figure 6 (b), removing the dielectric layer and using a 0.35 μm CMOS process to realize simple structure, small size, good performance and so on. The effective dielectric constant up to 48 and the quality factor reaches 20-40 in range 10 GHz from to 40 GHz.

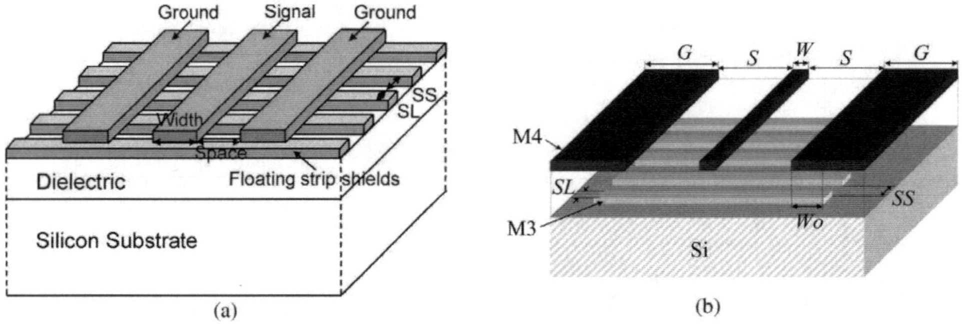

Figure 6. (a) Slow-wave CPW structure with gap-type floating protection, (b) Simplified form.

2.4. Research on modeling method of inductance

In the MMW-CMOS IC, inductors are widely used in low-noise amplifier (LNA), mixer, voltage-controlled oscillator (VCO), filter and other devices. With the development of CMOS technology and shrinking area of chip, requiring the inductance turn smaller, external inductors or active inductors are difficult to be realized. Because the traditional spiral inductor has big area and higher power consumption, which affect the performance of the various components of the integrated circuit, so the designing high self-resonant frequency (SRF) and high-quality factor Q inductance become essential task. Cao *et al.* [25] developed a differential circular helix inductance based on the traditional spiral inductor structure by using a 0.13 μm CMOS process, as shown in Figure 7, using M1 and POLY to make graphics shield (PGS) and adjusting diameter D to get different inductance and Q value.With development of BiCMOS process, has attract great attention, in 2017, Dickson *et al.* [26] proposed usage stacked inductor with compact inductor model. Measured results of the first dc to 50 GHz broad-band mixer and 40 GHz VCO with integrated inductors verifies the applicability of spiral structures and the Gilbert cell employed in the broad-band mixer demonstrates for the first time the benefits of a true BiCMOS topology for high-linearity, low-noise, and low-power millimeter-wave circuits.

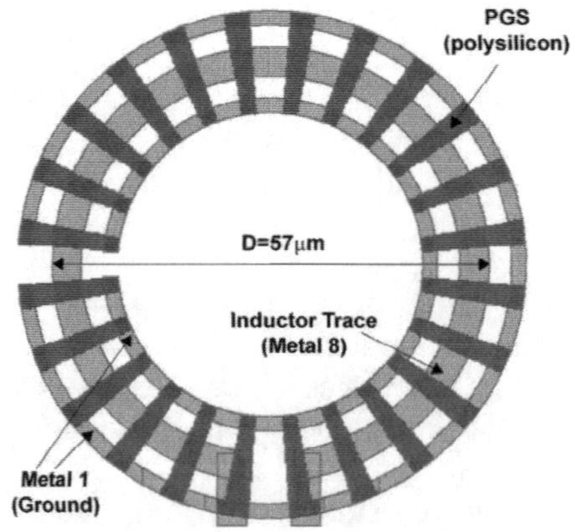

Figure 7. Differential circular spiral inductor structure.

3. Discussion and conclusion

In summary, the current MMW-CMOS IC technology still has the following key issues:

First, in the aspect of modeling to establish the de-embedding research. As the operating frequency of integrated circuit devices constantly increase, it is difficult to accurately characterize fixture parameters with simple equivalent circuit structures. In the modeling of MMW-CMOS IC, the traditional de-embedding method should be improved to accommodate the new MMW-CMOS IC structure.

Second, the traditional CMOS reduced model is only suitable for low frequency, when the operating frequency reaches the millimeter wave frequency, the simulation will appear to big error. In the modeling, we must first consider the CMOS parasitic parameters under high-frequency, and then researching model, using the software to extract the parameters of the CMOS transistor, finally complete layout to verify the design.

Third, in the aspect of researching transmission line. Because the traditional modeling principles and CMOS process have different feature, it is obviously initial that utilize standard CMOS process (substrate for the silicon) transmission line modeling. At present, the research on transmission line modeling is mainly aimed at transmission line and CPW research. Although it has the characteristics of small size and high frequency bandwidth, its substrate loss is considerable, the researchers proposed a passivation layer with a thin film transmission line and it reduce substrate loss, however, which increased the difficulty of process, cost and Q value. Due to the CPW easily connecting active or passive devices, this can achieve high Q values that widely used in millimeter-wave integrated circuits. The optimization of CPW modeling is mainly based on the gap-type floating protection technology and propagation wavelength, attenuation loss and characteristic impedance of CPW can be adjust by changing the length of the protective layer, the belt spacing and the protective layer. How we reduce the loss, the complexity of design and improve the Q value still is the focus of the research.

With the size of CMOS process constantly decreasing, the operating frequency of CMOS integrated circuit continues to be improved, MMW-CMOS IC will be moving towards the chip system and MMIC development, making the millimeter-wave baseband processor, transceiver and other modules integrated in a chip be possible. Although there are still many problems with the key technologies of MMW-CMOS IC, the related researching is not yet perfect and CMOS modeling techniques for active and passive devices need to be improve.With the continuous improvement of some key technologies and nanometer CMOS technology is turning mature, we always believe that the CMOS integrated circuit transceiver used in millimeter-wave wireless communication will be developed.

4. Acknowledgment

This work was supported by the National Natural Science Foundation of China under Grants 61504013 and 61702052, and by the Scientific Research Fund of Hunan Provincial Education Department under Grant 15C0033, and the Open Research Fund of Hunan Provincial Key Laboratory of Intelligent Processing of Big Data on Transportation, China.

5. References

[1] Cao J 2014 *Doctoral Dissertation* Southeast University (in Chinese).
[2] Yu F, Li P and Wang C 2016 *Solid State Electronics Research and Progress* 36 195 (in Chinese)
[3] Zhu F 2014 *Doctoral Dissertation* Southeast University (in Chinese).
[4] Stumper U and Schrader T 2014 *IEEE Trans. Instrum. Meas.* 63 1648
[5] Ding K, Yu Y and Zhou H 2015 *Journal of Nanjing University of Posts and Telecommunications (Natural Science Edition)* 35 40 (in Chinese)
[6] Wan L, Li Q, Wang Z, *et al.* 2015 *IEEE Trans. Instrum. Meas.* 64 694
[7] Prajapati H and Shah V 2014 *2014 Fourth Int. Conf. on Advanced Computing and Communication Technologies (ACCT)* P 335
[8] Kolding T 2000 *IEEE Trans. Electron Devices* 47 734
[9] Rosales M, Alarcon L and Sabido D 2006 *TENCON 2006-2006 IEEE Region 10 Conf.* P 1
[10] Niu G and Wei X2008 *9th Int. Conf. on Solid-State and Integrated-Circuit Tech.* P 472
[11] Issakov V, Wojnowski M, Thiede A, *et al.* 2009 *In European Microw. Conf.* P 695

[12] Kang I, Jung S, Choi T, *et al.* 2009 *IEEE Electron Device Letters* 30 398
[13] Doan C, Emami S, Niknejad A, *et al.* 2005 *IEEE J. solid-state circuits* 40 144
[14] Li S 2006 *Master's Thesis* Shanghai University (in Chinese).
[15] Hsu H, Chang J, Su J, *et al.* 2001 *IEEE MTT-S Int. Microw. Symposium Digest* P 1869
[16] Zhu Y, Wang W and Xia L 2007 *Microelectronics* 37 151
[17] Liang C and Razavi B 2009 *IEEE J. Solid-State Circuits* 44 450
[18] Wu H, Weng M, Su Y, *et al.* 2007 *IEEE Microw. Wireless Compon. Lett.* 17 673
[19] Lee S, Lee J, Kim Y, *et al.* 2007 *IEEE Trans. Compon. Packag. Technol.* 30 849
[20] Chiang M, Wu H and Tzuang C 2007 *IEEE Trans. Microw. Theory Tech.* 55 2512
[21] Sang L and Chan M 2015 *Int. Conf. on Infrared, Millimeter, and Terahertz Waves* P 1
[22] Zhu H, Xue Q, Liao S, *et al.* 2017 *IEEE Trans. Microw. Theory Tech.* 65 119
[23] Cho H, Yeh T, Liu S, *et al.* 2009 *IEEE Trans. Electron Devices* 56 1705
[24] Kaddour D, Issa H, Franc A, *et al.* 2009 *IEEE Microw. Wireless Compon. Lett.* 19 542
[25] Cao C and O K K 2006 *IEEE J. Solid-State Circuits* 41 1297
[26] Dickson T, Lacroix M, Boret S, *et al.* 2005 *IEEE Trans. Microw. Theory Tech.* 53 123

IOP Conf. Series: Materials Science and Engineering **351** (2018) 012012 doi:10.1088/1757-899X/351/1/012012

Qualitative study on maintenance management in Moroccan industries

Amal NAJI, Mohamed EL OUMAMI, Otmane BOUKSOUR and Zitouni BEIDOURI

Laboratoire de Mécanique, Productique et Génie Industriel,
Ecole Supérieure de Technologie, Hassan II University of Casablanca, Morocco

naji.amal@gmail.com; mohoumami@gmail.com; bouksour2@gmail.com; zbeidouri@gmail.com;

Abstract. Maintenance management is, and has been studied in depth, especially for strategies to be implemented in industry, even though, authors note that there is a gap between literature and management adopted by industries. In this paper, we present a qualitative study in five Moroccan industries to investigate "how maintenance is managed" rather than "how it should be". The questionnaire utilized for the study consists on semi-structured and open-ended questions. We consider factors and variables related to maintenance management and we explore the relationships between those factors. The original contribution of this paper is to provide a « real view » about maintenance management in Moroccan industries, which could help to improve understanding of barriers to implementing maintenance strategy.

1. Introduction

Maintenance is not a function that stand alone in a company; it is a mix of management, operations, technology and business strategies, [1]. Due to the « interdisciplinary cross functional nature » of maintenance, its management is a part of a broad area of the company organization [2]. For Ostadi [3], maintenance management system (MMS), is a set of technical, administrative and management activities during the life cycle of equipment that aim to improve condition and functioning to an expected level.

When companies decide to improve the investment in maintenance management, it is often a challenge, as it is not a straightforward decision [4]. Therefore, research in maintenance should consider all the factors related on and explore in-depth the relationships between those factors.

As an emerging country, Morocco orients its political economy to increase industrialization by developing industrial zones and encouraging foreign investments. Industry in Morocco, contributes in about one-third of the annual GDP (Gross Domestic Product) and the main industries in Morocco are Phosphates, rock mining and processing, automotive and aviation, food processing, leather goods, textiles and construction.

Maintenance is an important issue for all these industries. That's why we took an interest in maintenance management in our country.

The purpose of this qualitative study is to explore in-depth maintenance management in five Moroccan industries by responding to two questions: What characterize maintenance management in those industries? What are the impact factors and the challenges to maintenance function?

In the first section, we present the method used to conduct the qualitative study. In the second one, we discuss results of the study followed by conclusions and limitations of our study.

2. Methods

2.1. Research model

The present study is a qualitative research, where the aim objective is to explore the variables that influence maintenance management, and to investigate the relationships between those variables in Moroccan industries. As cited by Creswell [5], the qualitative research is used in order to explore a problem or a subject. For Gerede [2], the objective in qualitative research is to collect in-depth data from people and events related to the topic of the research.

Since, maintenance management depends on measurable and non-measurable variables, it is necessary to use search tools like direct observation of events and interview with people involved in this issue. According to Yin [6], case studies are the preferred method when the objective of study is to answer questions like "how" and "why". In this study, we select five case studies in industry.

2.2. Participants

The study is limited to five industries in Morocco. The participants are managers who set or implement maintenance strategies in their companies. The demographic data of the participants is shown in Table 1.

The five industries are large companies located in Morocco as detailed in Table 2 and they present similarities:

- Companies are founded for more than 20 years, (company E is founded in Europe for more than 20 years and its subsidiary is located in Morocco 10 years ago)
- The lifetime of the companies' equipment varies between 20 to 40 years, and is highly costly
- Their businesses are in a competitive environment
- Their work environment is risky.

Table 1. The demographic data of the participants

Gender			**Male**			**Female**
			5			
Age	**25 - 30**	**30-40**	**40-50**	**50-60**		
	1	2	2			
Education	**Under graduated**	**Master's / Engineer**	**PhD**			
		5				
Position	**Maintenance manager**	**Maintenance Director**	**Equipment Manager**	**Operation Manager**	**Site Director**	
	1	1	1	1	1	
Experience in the position	**Less than 5 years**	**5 - 10 years**	**10 - 15 years**	**15 - 20 years**	**More than 20 years**	
	1	2		2		

2.3. Data collecting tool

The literature associated to empirical studies on maintenance management and factors defined throw visits to industries were used to prepare the questionnaire. The development framework of the questionnaire is presented in Fig. 1.

The final version of the questionnaire comprises two parts. The first part of the questionnaire includes 77 semi-structured questions that are classified by subcategories related to variables and success factors of maintenance management. The second part consist of three open-ended questions in order to enable the participants to express freely theirs opinions about the following points:

- Maintenance management status in the company as perceived by them
- The barriers and the challenges for implementation of an optimal maintenance management strategy
- Added useful information about maintenance that was not included in the questionnaire or during interviews.

In order to collect data, we conducted face-to-face interviews with participants, in addition to exchanges via mails and conference calls while analyzing results.

The interviews consist on three parts;

1) We present the objective and the benefits of the study, and we ask the responder to present the company, the activity, products or services provided to clients, processes, equipment and the organizational chart.
2) The participant answer to the first part of the questionnaire, which consists of numerical and semi-structured questions.
3) The participant answers to the second part of the questionnaire, which consists of three open-ended questions. During this section, we discuss profoundly the impact of organization culture and the constraints that managers face to implement maintenance strategy.

2.4. Data analysis and interpretation

Qualitative research requires collecting data from the context where events occur in order to describe this occurrence. The aim objectives of qualitative research are describing, decoding and understanding eclectic data, [7].

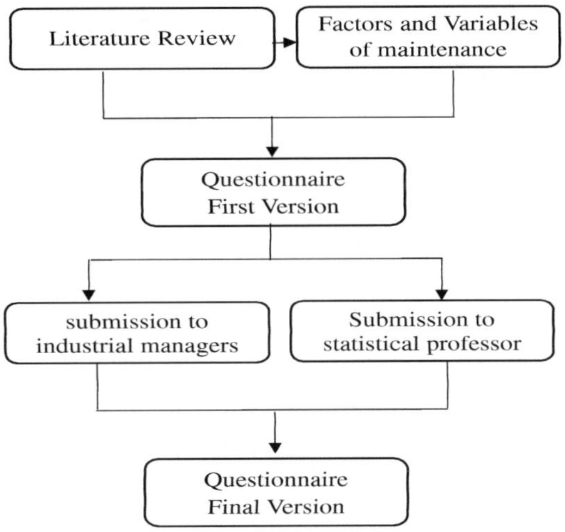

Figure 1. The development framework of the questionnaire

Table 2. Sectors and activities of the five companies studied

Company	Sector	Activities	Work hours
A	Logistic	Port operator	24h/24
B	Cement	Cement production	24h/24
C	Energy	Gas production	2 shift / day (16h)
D	Agri food	Oils production	24h/24
E	Logistic	Maritime transport	24h/24

The information collected was compiled. For each company, we evaluated separately each of subcategories. Then, we analyzed relationships between factors and variables to explain the causality of issues. Due to the complexity of the maintenance function, we contacted the responders in order to validate assumptions and interpretations. Finally, we compared the research findings to results in literature.

3. Results and discussion

3.1. Subcategories and degree of importance

Among the subcategories of the questionnaire, factors influencing maintenance management in the five industries studied were discovered through the qualitative analysis of results as shown in the figure 2. We notice that top management, maintenance department and human resources are the most impacting factors. In the following sections, we will discuss each of those defined factors and we will investigate the challenges and barriers to the maintenance management implementation.

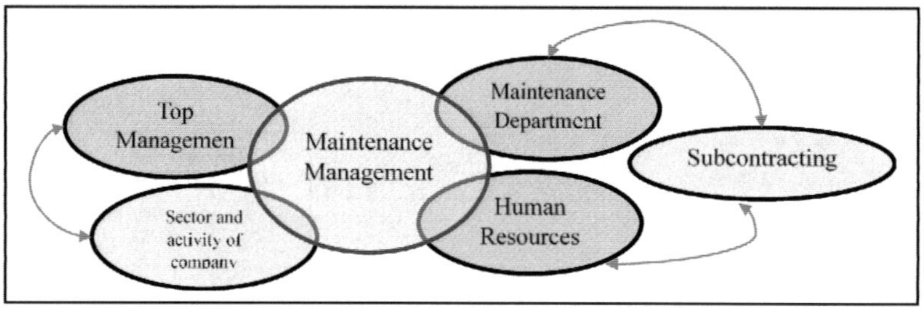

Figure 2. Factors influencing maintenance management in the five industries studied

3.2. The impact of top management policy

Despite the procedures and objectives set by top management, it is the culture of the company that define the policy to be applied and adopted by people. Participants of companies A, C and D consider that availability and safety of equipment are more important than safety of staff, even though, companies organized periodically training session about safety personnel. Bupe Mwanza [8] considers that safety during maintenance works is also important and could save additional cost such as, worker's compensation, health insurance claims, regulatory fines, lawsuits and so many more.

The top management of the four companies A, B, C and E has an annual objectives but not a long term vision among one year, their priorities are essentially availability of equipment, respect of the budgets and maintenance preventive rates, therefore, no company have implemented a PMS (Performance Maintenance System). The process of maintenance performance measurement (MPM) is derived from the overall organizational strategy and should be oriented by the integration of critical success factors (CSF), ([9]–[11]) . The participant of company A said that currently, the implementation of measures such MTBF, MTTR and LCC are ongoing in order to improve the maintenance management. In other hand, the participant of company C explained that the utilization of MTBF in the past, did not

improve the maintenance management, he believes that improvement of maintenance function must be more profound than calculating measures. Generally, the petroleum industry use the asset management approach that include the technical and the business aspects of an asset in a complex sociotechnical environment between organizations, personnel, technology and systems, [12].
We notice that consequences of failures have a crucial impact on the management maintenance policy. For company E, a failure could have catastrophic consequences, therefore top management set procedures and implement measures to improve continuously the maintenance function. According to Lazakis [13] , the preventive maintenance is the best approach for shipboard equipment. For the company D, top management do not set annual objectives, as there are four lines production, while three lines are sufficient to satisfy the production rate required. However, the company had a production shutdown in a production line due to an accidental breakdown. Currently, the company are reviewing the organization and policies to remedy to the situation, which causes decrease of profitability ratio.

3.3. Challenges of maintenance department

Maintenance department set maintenance strategy and annual planning according to the top management objectives. Four companies have a preventive maintenance strategy as detailed in the Table 3. When setting the planning of maintenance operations, the managers consider experience feedback more than numerical measures. This is due to the lack of statistics, as most companies have an overall statistics about the failure rate and not detailed information about it. This issue is cited by Gopalakrishnan [14], with reference to his survey, he found a gap between criticality classification and maintenance planning industries, where bottleneck detection is widely used.
In our study, company B, uses RCA (Root Cause Analysis) to analyse fatal failure; we notice that the annual maintenance budget of this company is ten times greater than other companies' budget.
Historical events and accidents contribute to the evolution of maintenance in industries, according to participant of company B; there has been fatal accidents in the site 15 years ago and then since that time, safety during maintenance operations is a priority for top management.

Table 3. Maintenance strategy in the five industries

Companies	Preventive Maintenance	Conditional Maintenance	Corrective Maintenance
A	80%		20%
B	70%	10%	20%
C	75%	5%	20%
D	10%	5%	85%
E	77%	11%	12%

The subcontracted maintenance is a policy adopted by four companies. We present in Table 4 the rates estimated by participants concerning subcontracting complex maintenance operations. For the company D, the maintenance management are supervised by a centralized department and all operations are executed by the maintenance staff, expect a limited complex operations. According to participants of companies C and D, subcontracting is a policy to face union, rather than maintenance strategy. We mentioned that Marttonen [15] specifies requirements for maintenance contract-level, including technical, organizational and legal arrangements from operative maintenance of single assets to strategic management.
We note that the spare parts management is neglected by all the five companies. The spare parts stored are generally overestimated; more than 50% of spare parts stored have not been used yet for more than one year. For participants of companies A and C, the spare parts are overestimated in order to avoid a shutdown of production as importation of critical spare part take above one month. Cholasuke [16], mentioned that an effective spare part management could be a significant source of cost saving. We notice that all participant of companies A, B and C agree that spare parts management is a weakness that should be improved.

We conclude that in large companies studied, challenges on maintenance are relating essentially to maintenance managers, as today top management consider maintenance as an important function by including it in the annual objectives. Smith [17] cited this issue, as he considers that currently, maintenance and asset managers have a good opportunity to be a part of the business view and that they should make a choice to be a leadership that will take a step forward and will positively influence in the organization.

3.4. The impact of human resources management on maintenance function

Managers generally agree that maintenance staff is a pillar for implementing maintenance strategy. Smith [17] go further by suggesting that maintenance managers should have a new approach of leadership by considering hard (plant) and soft (people) assets.

In our study, we have devoted a large section to human resources management. Sheikhalishahi [18], presented an organization of publications related to human factors in maintenance into three main categories : human error / reliability calculation methods, workplace design / macro-ergonomics and human resource management. Ahmadzadeh [19], present a model to prioritize maintenance-related waste caused by human factors. During interviews, we deduce that such classification are not used in companies surveyed, as measurement of human factors are not yet applied.

Table 4. Subcontracting rate in the five industries

Companies	Percentage of subcontracted complex maintenance operations	Percentage of subcontracting cost Vs Maintenance cost
A	90%	70%
B	50%	30 to 40%
C	80%	60%
D	70%	70%
E	15%	11%

Table 5. Documentation of maintenance operation for operators

Companies	Scheduled maintenance tasks	Tool to be used	Measures before and after the operation	Safety measures	Duration of the operation
A	x				
B		x	x	x	
C			x	x	x
D					
E			x		x

The maintenance staff in the five companies has mainly the same demographic structure; they are composed of two categories in accordance to their age:

- Elderly staff, they have more than 10 years of experience in maintenance service, and are not graduated
- Young staff, they have generally five years of experience and are undergraduate, in maintenance or in production.

The most important human performance factors are 'motivation' and 'competence' [20]. We include those factors in the questionnaire. For motivation factor, the participants of all companies confirm that operators execute frequently risky tasks, but only company A and B provide to the personnel bonuses for risky tasks. For competence factor, participants of companies A, B, C and D affirm that lack of technical skills of maintenance staff constitute a barrier for improving maintenance management. According to the participants of companies A, B and C, it is due to the complexity of the equipment, which require a continuous training session and profound analysis to maintain it. However, we notice that maintenance operations are not totally documented, to guarantee the accomplishment of maintenance task as shown in Table 5.

Also, training session are not sufficient regarding the complexity of tasks and of machine technologies either for maintenance managers who inform us that they ensure technologies watch by learning from subcontractors.

In parallel to the lack of technical knowledge and skills evoked by participants, we remarked that all companies subcontracted most of the maintenance operations. We conclude that subcontracting most of maintenance operations may contribute to transforming the role of maintenance department to a management supervisor.

3.5. Challenges and barriers to maintenance management implementation

We present challenges and barriers to implementing a maintenance strategy in Moroccan industries according to the five participant's response:

- Top management policy: four companies have written procedures and a clear organizational chart of maintenance department, expect for the company D. However, the culture of organization is more important and may vary from written procedures and objectives.
- Lack of technical skills and knowledge: this obstacle concerns operators and managers. The participant of company A note that accidents may occur when a manager do not analyze the equipment operation. The participant of company C think that managers should have skills in order to control maintenance operations executed by subcontractors
- Complexity of machine and equipment: industries are facing challenges like optimization of maintenance function due to the evolution of technologies, [21]. The participant of company A think that it is difficult to ensure a technology watch on port equipment due to the continual evolving of technologies; also, participant of company B, evoke that the company failed to decentralize maintenance department, due to lack of technical skills.
- The production workhours: participant of company E evoke that time available is insufficient to accomplish preventive maintenance, because it is done during the stopover of ship. The participant of company C explained that the preventive maintenance operations are planned every Saturday, during the stoppage of the production, but if there is a need to increase the production rate, maintenance tasks are postponed to Sunday, or reduced to twice a month.

4. Conclusions and limitations of the study

We discovered through this qualitative study that in large companies, where preventive maintenance is up to 80%; production and top management policy are no more the important barriers for implementing maintenance strategy. Today, to improve maintenance function, challenges are related to technical skills and knowledge of maintenance staff. Therefore, maintenance managers should move from the day-to-day management of operations to long term vision by suggesting training session, diagnosing failures and analyzing equipment operations. As mentioned by Macchi and Fumagalli [22], a company should classify the criticalities in its maintenance processes and make a benchmark with the best companies to better drive the investment. This will help top management to view maintenance management in a strategic perspective and will contribute to set mid and long-term goals.

This qualitative study focused on maintenance management on large companies, where maintenance is considered as an important function. Companies are located in Casablanca and Tangier, which are industrial cities in Morocco, then we could not extrapolate the finding of this research to small and medium companies in small cities. We consider complementing this qualitative study as mentioned by Yin [6], by "exploratory" and "descriptive" case studies in industries before generalizing the finding of this study.

References

[1] L. Pintelon and A. Parodi-Herz, "Maintenance: an evolutionary perspective," in *Complex system maintenance handbook*, Springer, 2008, pp. 21–48.

[2] E. Gerede, "A study of challenges to the success of the safety management system in aircraft maintenance organizations in Turkey," *Saf. Sci.*, vol. 73, pp. 106–116, 2015.

[3] B. Ostadi and H. Saifpanahi, "A Practical Self-Assessment Framework for Evaluation of Maintenance Management System based on RAMS Model and Maintenance Standards," *J. Ind. Syst. Eng.*, vol. 10, pp. 125–143, 2017.
[4] L. Pintelon, S. K. Pinjala, and A. Vereecke, "Evaluating the effectiveness of maintenance strategies," *J. Qual. Maint. Eng.*, vol. 12, no. 1, pp. 7–20, Jan. 2006.
[5] J. W. Creswell, *Research design: Qualitative, quantitative, and mixed methods approaches*. Sage publications, 2013.
[6] R. K. Yin, *Case study research: Design and methods*. Sage publications, 2013.
[7] M. Hlady-Rispal and E. Jouison-Laffitte, "Qualitative research methods and epistemological frameworks: A review of publication trends in entrepreneurship," *J. Small Bus. Manag.*, vol. 52, no. 4, pp. 594–614, 2014.
[8] B. G. Mwanza and C. Mbohwa, "Safety in Maintenance: An Improvement Framework," *Procedia Manuf.*, vol. 8, pp. 657–664, 2017.
[9] A. Parida, U. Kumar, D. Galar, and C. Stenström, "Performance measurement and management for maintenance: a literature review," *J. Qual. Maint. Eng.*, vol. 21, no. 1, pp. 2–33, Mar. 2015.
[10] J. K. Leidecker and A. V. Bruno, "Identifying and using critical success factors," *Long Range Plann.*, vol. 17, no. 1, pp. 23–32, 1984.
[11] A. H. C. Tsang, A. K. S. Jardine, and H. Kolodny, "Measuring maintenance performance: a holistic approach," *Int. J. Oper. Prod. Manag.*, vol. 19, no. 7, pp. 691–715, Jul. 1999.
[12] M. Kusumawardhani, M. Kusumawardhani, S. Gundersen, S. Gundersen, M. Tore, and M. Tore, "Mapping the research approach of asset management studies in the petroleum industry," *J. Qual. Maint. Eng.*, vol. 23, no. 1, pp. 57–70, 2017.
[13] I. Lazakis and A. Ölçer, "Selection of the best maintenance approach in the maritime industry under fuzzy multiple attributive group decision-making environment," *Proc. Inst. Mech. Eng. Part M J. Eng. Marit. Environ.*, vol. 230, no. 2, pp. 297–309, 2016.
[14] M. Gopalakrishnan, J. Bokrantz, T. Ylipää, and A. Skoogh, "Planning of maintenance activities–A current state mapping in industry," *Procedia CIRP*, vol. 30, pp. 480–485, 2015.
[15] L. Metso, S. Marttonen, N. E. Thenent, and L. B. Newnes, "Adapting the SHEL model in investigating industrial maintenance," *J. Qual. Maint. Eng.*, vol. 22, no. 1, pp. 62–80, 2016.
[16] C. Cholasuke, R. Bhardwa, and J. Antony, "The status of maintenance management in UK manufacturing organisations: results from a pilot survey," *J. Qual. Maint. Eng.*, vol. 10, no. 1, pp. 5–15, Mar. 2004.
[17] B. J. Smith, "Optimising the Maintenance Function-It's Just as Much About the People as the Technical Solution," in *Engineering Asset Management*, Springer, 2006, pp. 568–575.
[18] M. Sheikhalishahi, L. Pintelon, and A. Azadeh, "Human factors in maintenance: a review," *J. Qual. Maint. Eng.*, vol. 22, no. 3, pp. 218–237, 2016.
[19] F. Ahmadzadeh and M. Bengtsson, "Using evidential reasoning approach for prioritization of maintenance-related waste caused by human factors—a case study," *Int. J. Adv. Manuf. Technol.*, vol. 90, no. 9–12, pp. 2761–2775, 2017.
[20] R. Peach, H. Ellis, and J. K. Visser, "A maintenance performance measurement framework that includes maintenance human factors: a case study from the electricity transmission industry," *South Afr. J. Ind. Eng.*, vol. 27, no. 2, pp. 177–189, 2016.
[21] R. S. Velmurugan and T. Dhingra, "Maintenance strategy selection and its impact in maintenance function: A conceptual framework," *Int. J. Oper. Prod. Manag.*, vol. 35, no. 12, pp. 1622–1661, Dec. 2015.
[22] M. Macchi and L. Fumagalli, "A maintenance maturity assessment method for the manufacturing industry," *J. Qual. Maint. Eng.*, vol. 19, no. 3, pp. 295–315, Aug. 2013.

Maintenance of buildings: Italian examples of deviations between planned and incurred costs

Manuela Rebaudengo[1] and Paolo Piantanida[2]

[1]Politecnico di Torino- DIST, Viale Mattioli 39, 10125 Torino, Italy
[2]Politecnico di Torino - DISEG, Corso Duca degli Abruzzi 24, 10125 Torino, Italy

manuela.rebaudengo@polito.it

Abstract. In recent years, drawing up the maintenance plan for a building has proved to be a theoretical practice, which has been expressed in the indication of a multiplicity of interventions more for caution reasons than for real needs. In this way, the document (mandatory by the Italian law both for public and private works) has been interpreted as a formal requirement: it must be part of the project, without any particular operational interest. So, the maintenance plans are more often "standard plans", not directing the user towards a specific maintenance strategy and thus consolidating the common thesis that reality is different. The paper compares, with reference to two xxx of study (offices and residence), the planned maintenance actions and those carried out, with an indication of the difference between expected and incurred costs.

1. Introduction

The Maintenance is defined as a complex of activities necessary to preserve the performance of technological and structural elements over time [1], with the aim of conserving the building assets (and also equipment) for the entire service life, that period of time between the service kick-off and the moment when the repair is uneconomic.
The Italian reference legislation (the new Code of Contracts, the Construction Act, UNI standards) clearly distinguishes between ordinary and extraordinary maintenance (Table 1).
While the New Code on Public Contracts and Buildings Consolidated Law indications are mandatory, the technical standards issued by UNI (Italian National Unification Body) are representative of a voluntary level and therefore may not be applied.

Table 1. Summary of national definitions.

	Ordinary Maintenance	*Extraordinary Maintenance*
New Code on Contracts [2]	*Repair, renovation and replacement works necessary to eliminate the deterioration of the structures, in order to preserve the condition and the usability of all the components, systems and*	*Works and modifications necessary to renew and replace structural and other parts of the buildings and the relative appliances, to adapt the components, systems and works connected with their use and the regulations in force and with the*

	connected works, keeping them in good working order and in safe conditions	*aim of reducing the significant deterioration*
Buildings Consolidated Law [3]	*Building works relating to the repair, renovation and replacement of building finishes and those necessary to integrate or maintain existing technological installations*	*Works and modifications necessary to renovate and replace structural and other parts of buildings, as well as to implement and integrate sanitary and technological facilities, provided that they do not alter the overall volume of buildings and do not entail any change in their intended use*
UNI Technical Standards [4,5]	*Minor repairs, requiring only small parts, which involves the use of consumables of normal use or the replacement of small parts*	*Particularly important activities or special equipment or instrumentation, which involves repairs and/or replacement parts, provides for the overhaul of structural elements*

Theoretical parallels between maintenance in construction and industry are frequent although in the first case we talk mainly of degradation and obsolescence, while in the second case of failure and the industrial outcome can usually lay on a robust reliability database concerning the behavior of serial products, while a building behaves as a prototype mainly.

During the life cycle of a building, maintenance is responsible for bringing the system's performance back to values that allow its use, despite the ageing of the asset; in other words, maintenance actions aim to compensate the "natural" decrease in the asset value and initial performance. As is well known, a building system is gradually declining and loses part of its functionality: this phenomenon is defined as obsolescence and can be limited but not eliminated, since, as in the case of building materials, it concerns its natural aging and loss of performance [6, 7].

From the theoretical point of view obsolescence can be normative, physical, technological, functional and formal. The most frequent causes may be mismanagement or inappropriate use; accidental phenomena or environmental effects; design or construction errors; the appearance of defects or abnormal behavior [7, 8].

Maintenance is related to different phases of the construction process: at first, to the design phase, when it is necessary to foresee it in technical and cost terms; to the construction ones, to check that it is carried out correctly; during the asset management, through minimum or more substantial interventions; at precise control times, when it is possible to think about possible intervention strategies.

Other definitions of maintenance [9, 10] refer to the intervention strategy: scheduled maintenance is, together with condition maintenance, a type of preventive maintenance; instead, corrective (or fault) maintenance is a type of maintenance that provides for operations to be carried out only after failure has been detected; the latter is a strategy that can be adopted where there are non-critical and easily substitutable systems at low cost: it is therefore more convenient to wait for the maintenance to be carried out.

These two approaches essentially distinguish the theoretical case (planned maintenance) from the real one (corrective maintenance): the paper will reveal, in both case studies, substantial cost differences, which will be useful in improving preventive estimates. The calculation of total costs in the theoretical hypothesis (indications of maintenance plans) will be limited only to those interventions with a high probability of being carried out, for example skimming a large part of preventive inspections. We expect different values, significantly lower than the theoretical ones; then we will compared them with the annual incurred costs, after a short and limited period of management.

2. Maintenance plan: structure and contents

The P D no. 207/2010, now overtaken with many critical issues by the New Code of Contracts [11, 12], defined the maintenance plan as a complementary document to the executive project, which includes graphical drawings and provides for the maintenance of the intervention with the aim of maintaining its functionality, efficiency and quality characteristics over time.

It consists of a set of documents (User's Manual, Maintenance Manual and Maintenance Program) which are intended to provide useful information for the correct use of the asset, its correct maintenance and management. The User Manual is the document that encloses the operating instructions in non-specialist language and is intended for users; the Maintenance Manual is intended for technical personnel who will arrange the maintenance or management of the property, therefore it will be written with a technical - specialist language. Finally, the maintenance programme, which is more operational, is divided into three sub-programme: of services, of controls and of operations.

The main weakness comes out from the scheduled step to provide all the maintenance documents. Before the bidding and the end of construction phase it's impossible to determine the manufacturer and the model of components provided by the contractor (during the construction step) and so part of the maintenance planning is necessarily generic (and thus is why the maintenance plan is considered as another bureaucratic requirement not linked to any useful operational issues).

The most commonly used UNI maintenance standards are UNI- EN 10604:1997 " Maintenance - Criteria For Design, Management And Control Of The Maintenance Services Of Building ", aimed at providing general management criteria and tools for the maintenance of real estate assets, and UNI 11257:2007 "Maintenance of real estate assets - criteria for drawing up the plan and maintenance programme of real estate - Guidelines", which details the criteria through which plans can be drawn up.

In operational terms, if there were buildings thematic archives showing the frequency of failures and pathologies, writing these documents would be easier, less repetitive and more accurate, and it would have more scientific validity than today. The most rigorous professionals refer to foreign databases, which nevertheless show critical issues linked to the reference compass and the construction technologies they represent.

3. Estimated maintenance costs for case studies

The analysis in this paper will demonstrate the necessity to overcome the "simplified methodology" provided by Italian law to plan the building maintenance, because of its unreliable connection with the real maintenance costs, both in residential and office buildings. To give evidence to these facts, we'll consider two examples, one residential and one office building. In both cases, these are new constructions with reinforced concrete load-bearing structure and vertical closing elements in brick, external plaster finish.

Table 2. Case studies: key figures summary.

	Residential Buildings	*Office Buildings*
Construction costs	*6,4Million Euros*	*3,77 Million Euros*
Year of construction	*2012*	*2016*
Area	*4.350 sq. m.*	*2.500 sq. m.*

In both cases, the maintenance programme is available, with an indication of the planned project of controls and interventions: as far as the residence is concerned, since it has been in operation for more than 5 years, the final accounts of the costs incurred for ordinary maintenance are also available.

The analysis approach aims to determine any deviations between theoretical and operational costs, derived from the final balance sheet. In particular, the research has been developed:

(1) by calculating the cost of checks and operations theoretically provided by the designers,

(2) by comparing it with the theoretical cost used for the preventive estimates;
(3) by identifying controls and operations that are most likely to be carried out (and their related costs)
(4) and by verifying the costs really incurred.

At first we defined inspections and intervention costs (provided by inspection and operation programme) through a quantity survey, where a typical workers team and an average time for each operation were estimated and these values were associated with the hourly labor costs (referred to the second half of 2016), with particular reference to the case studies geographical area (Piedmont Region) [13]. The overall value, appears to be largely overestimated, because the controls provide for visual investigation of all the technological elements that can be inspected and the interventions for the restoration of all these elements, in the event of failure or pathology that compromises their correct use. The first comparison wants to relate these theoretical design activities/costs (which are defined in the executive design phase) and the synthetic forecast ones, typical of the programming phase. The value, so far from the initial forecasts, depends on maintenance plan contents, often related to future actions and containing even more than the actions normally due...

We simply considered the control activities propensity (in Italy, it is not usually done in a preventive way, but in a "corrective"); therefore, we choose to consider only controls over those technological elements with a higher probability of failure was considered.

In order to determine this probability of occurrence, some foreign databases were consulted and the information was verified through market surveys of Italian operators in the construction sector. Then we decided to estimate the costs of inspections for which the occurrence rate was more than 25% in order to increase the reliability of this survey methodology. As far as the interventions are concerned, on the other hand, and in the same frame the costs of the interventions have been calculated, at which the probability of occurrence is greater than 30%. At the end, the actual costs are those derived from the items of expenditure actually incurred by the owners. Table 3 shows briefly the results.

Table 3. Summary of research results.

	Ordinary Maintenance (annual costs)	
	Residential Buildings	*Office Buildings*
Planned costs (from design phase)	*0,43% of construction costs*	*0,57% of construction costs*
Planned costs (from programming phase)	*0,30% of construction costs*	*0,30% of construction costs*
Planned costs (by occurrence analysis)	*0,18% of construction costs*	*0,18% of construction costs*
Real costs	*0,15% of construction costs*	*0,16% of construction costs*

4. Conclusions and future research developments

The processed data are certainly interesting, because they show how often people tend to overestimate these costs in the forecasting phase or prescribe behavior far from what happens in reality. However, some clarification is needed. The data collected can be used for the intended use indicated and for buildings of the same age and destination/technology. So these parameters cannot be used extensively on all existing assets. For this reason, it is necessary to deepen the theme by extending the data

IOP Conf. Series: Materials Science and Engineering **351** (2018) 012013 doi:10.1088/1757-899X/351/1/012013

collection to other destinations of its more typically public (sporting, social, health,...) or private (commercial, industrial,...) and to cases of intervention on existing heritage. From a theoretical point of view, for some categories of components, generations of older buildings (1995-2004) have fewer defects than younger generations (2010-2012), even if the latter do not yet have a significant burden of disease on the whole guaranteed life of the building; this observation can generally be extended to all comparisons between generations of constructions. The analysis of the databases, even if not directly referable to the Italian building heritage, therefore also with different construction techniques, provides interesting data but does not yet allow to obtain a frequency of defects on the various building components. In conclusion, the two investigated case studies show the provisional simplified design methodology for maintenance planning is very far from the real maintenance costs. For this reason, even it is admitted by Italian law, it should be avoided and substituted with a reliable and wide spread database of real life maintenance service for buildings.

5. References

[1] Marcon F., Re Cecconi F. (2012). Manutenzione e durata degli edifici e degli impianti. Maggioli Editore; ISBN / EAN 8838774919 / 9788838774911 (pg.5-11)

[2] L. D. no. 50/2016. Ordinary Supplement to GURI no. 10 of 19 April 2016

[3] P. D. no.380/2001. Ordinary Supplement to GURI no. 245 of 20 October 2001

[4] UNI- EN 10604:1997 " Maintenance - Criteria For Design, Management And Control Of The Maintenance Services Of Building"

[5] UNI 11257:2007 "Maintenance of real estate assets - criteria for drawing up the plan and maintenance programme of real estate - Guidelines"

[6] Roscelli R. a cura di, (2014), MANUALE DI ESTIMO. Valutazioni economiche ed esercizio della professione. Utet Università (pg.

[7] Campbell, J. D., & Reyes-Picknell, J. V. (2015). Uptime: Strategies for excellence in maintenance management. CRC Press.

[8] Martinelli S. (2016). Manutenzione edifici in condominio. Grafill (pg. 5-15)

[9] Ahmed E. Haroun, (2015) "Maintenance cost estimation: application of activity-based costing as a fair estimate method", Journal of Quality in Maintenance Engineering, Vol. 21 Issue: 3, pp.258-270, doi: 10.1108/JQME-04-2015-0015

[10] Lindley R. Higgins; R. Keith Mobley; Darrin Wikoff: Maintenance Engineering Handbook, Seventh Edition. ESTIMATING REPAIR AND MAINTENANCE COSTS, Chapter (McGraw-Hill Professional, 2008 2002 1995 1988 1977 1966 1957), AccessEngineering

[11] Prizzon F., Rebaudengo M. (2017), Il nuovo Codice dei contratti pubblici, tra inno-vazioni e criticità, Territorio Italia n.1/2017; ISSN 2499-2666

[12] Rebaudengo M., Prizzon F. (2017), Assessing the Investments Sustainability After the New Code on Public Contracts, in Computational Science and Its Applications - ICCSA 2017, 17th International Conference, Trieste, Italy, July 3-6, 2017, Proceedings, Part VI, Lecture Notes in Computer Science vol. no. 10409, pg. 473-484, ed. Springer, ISBN: 978-3-319-62406-8

[13] http://www.regione.piemonte.it/oopp/prezzario/dwd/index.htm

Acknowledgements

Special thanks to Elisa Boscolo, who worked on these topics during his Master Degree Thesis at Politecnico di Torino and to Luca Lacidogna, who collected some information on maintenance costs trends and related strategy.

IOP Conf. Series: Materials Science and Engineering **351** (2018) 012014 doi:10.1088/1757-899X/351/1/012014

A bi-objective model for optimizing replacement time of age and block policies with consideration of spare parts' availability

Imad Alsyouf

Industrial Engineering and Engineering Management Department, College of Engineering. Sustainable Engineering Asset Management (SEAM) Research Group, University of Sharjah, Sharjah, United Arab Emirates

ialsyouf@sharjah.ac.ae

Abstract. Reliability and availability of critical systems play an important role in achieving the stated objectives of engineering assets. Preventive replacement time affects the reliability of the components, thus the number of system failures encountered and its downtime expenses. On the other hand, spare parts inventory level is a very critical factor that affects the availability of the system. Usually, the decision maker has many conflicting objectives that should be considered simultaneously for the selection of the optimal maintenance policy. The purpose of this research was to develop a bi-objective model that will be used to determine the preventive replacement time for three maintenance policies (age, block good as new, block bad as old) with consideration of spare parts' availability. It was suggested to use a weighted comprehensive criterion method with two objectives, i.e. cost and availability. The model was tested with a typical numerical example. The results of the model demonstrated its effectiveness in enabling the decision maker to select the optimal maintenance policy under different scenarios and taking into account preferences with respect to contradicting objectives such as cost and availability.

1. Introduction

Reliability and availability of critical systems play an important role in achieving the stated objectives of engineering assets. Maintenance refers to the measures taken by the users of a product or a system to keep it in an operable condition, or to repair it in order to restore it to an operable condition [1]. Usually, these measures aim at minimizing failures and their consequences. Traditionally, maintenance is performed either due to a failure, i.e. corrective maintenance, or performed according to schedule to prevent a failure from happening, which is known as preventive maintenance. In both cases, there is a need to have spare parts, to replace the worn or faulty ones. If the spare parts are not available, the system's downtime is extended to a longer time, resulting in unnecessary expenses and maybe even safety risks. Spare parts' inventory level is a critical indicator that is affected by various factors such as previous demand, financial and technical asset management. Wang [2] illustrated that, generally, spare parts are managed through an inventory management system that relies on past spare parts demand regardless of the maintenance process. On the one hand, an insufficient stock level may lead to an extended system downtime. On the other hand, a high stock level results in higher holding costs. The spare parts' demand is usually affected by the maintenance policy. Preventive maintenance with less frequent scheduled maintenance leads to less used spare parts as well as to a higher probability of corrective maintenance's occurrence. Preventive maintenance with more frequent scheduled maintenance might lead to a higher demand of spare parts, but minimized risk of system's breakdown (see Wang [3]). It is usually assumed that spare parts are always available when needed, that is the focus is mainly set on maintenance. Wang [2] emphasized that maintenance scheduling and spare parts' management should be integrated to achieve the best results. He also cited publications [4–11] that address the mentioned integration. Further, publications that address both the optimization of maintenance scheduling and spare parts' management are mainly based either on an age replacement policy, such as [12,13] or on a block policy, such as [14–16]. Nosoohi and Hejazi [13] presented a multi-objective model to simultaneously determine the spare parts' number and preventive replacement times in a planning horizon. They used an age maintenance policy with three objective measures in their study, i.e. cost, corrective failure and residual lifetime. They cited research papers that considered maintenance policies based on system reliability and availability by using reliability and availability as objective measures [17–19].

In this paper, we will investigate the use of bi-objective model, i.e. cost and availability, to determine the optimal replacement time with and without spare parts availability for three preventive maintenance policies, i.e. age, block Good-As-New (GAN) and block Bad-As-Old (BAO). Compared to Nosoohi and Hejazi [13] and other research work, the main contribution of this paper is the consideration of not only of the age or block policies but also of the age, block Good-As-New (GAN) and block Bad-As-Old (BAO) policies, simultaneously.

The organization of the paper is as follows. Section 2 presents maintenance approaches. Section 3 illustrates the suggested model and its decision variables and parameters. Section 4 provides a numerical example to illustrate how the proposed model works. Finally, conclusions are provided in section 5.

2. Maintenance approaches

Maintenance approaches are classified as reactive or preventive. For more details on maintenance approaches and their development see [1,20,21]. Reactive maintenance can be either run-to-failure in cases where the failure has little or no effect or breakdown maintenance due to failures of critical components [1]. On the other hand, preventive maintenance is defined by [9] as "any task designed to prevent failures or mitigate their effects". Preventive maintenance can be either time-based or condition-based. With time-based policies, critical components are replaced at pre-specified intervals regardless of their conditions. The time for action is usually determined based on experience, suggestion by the original equipment manufacturer (OEM) or optimized by minimizing maintenance total costs using life distributions generated from failure data [1].

Age and block policies, which were first proposed by [7], are widely used for optimizing the renewal intervals for components. When using age policy renewal is done either at a predetermined fixed age (running or calendar time since its last renewal) or when it fails, whichever event occurs sooner. However, with block renewal policy the time of preventive renewal is never interrupted or changed due to failures that occur between the planned renewals. The cycle time is fixed and there may be any number of failures between the planned renewals. Using block policy one has two options, either to replace the component with a new one, i.e. restoring the item to a standard condition Good-As-New (GAN), or to repair the component with minimum repair, i.e. Bad-As-Old (BAO). These models differ in their impact on the hazard rate function. For example, with both age and block-GAN models the time variable in the hazard function is restored to zero, whereas with the block-BAO model the time variable in the hazard function is assumed not be affected. [8,9].

Condition based maintenance is a strategy where the decision to perform maintenance is reached by observing the "condition" of the system and/or its critical components. It is applied to items which fail due to aging or accidently [22] and implies the use of surveillance and monitoring techniques to monitor the condition of critical components [20,23].

3. Model

The proposed model considers both the unit cost and the availability to optimize the maintenance interval under three time-based preventive maintenance policies, namely age, block-GAN and block-BAO. The model considers the possibility of having available spare part in its formulation. This formulation represents a generalized model, where the model without available spare parts can be seen as special case as will be illustrated later.

3.1. Model parameters and decision variables

The optimization model parameters are defined as follows: C_{f1} and C_{f2} are the cost of fixing failure with and without the availability of spare parts and are used in the case of age and block-GAN models. C_{p1} and C_{p2} are the cost of performing preventive maintenance with and without spare parts. C_{r1} and C_{r2} are the costs of repairing failure with and without spare parts used in the case of block-BAO. In this case, the failed component is replaced by a reconditioned one or restored to its "as bad as old" state. T_{f1} and T_{f2} are the time required to fix failure with and without the availability of spare parts and are used in the case of age and block-GAN models. T_{p1} and T_{p2} are the time needed to perform preventive maintenance with and without spare parts respectively. T_{r1} and T_{r2} are the time required to repair failure with and without spare parts.

$P(t)$ defines the cumulative probability of having spare parts, $F(t)$ is the failure cumulative distribution function. $R(t)$ defines the component reliability function, $N(t)$ represents the expected number of failure in block-GAN model and $H(t)$ is the cumulative hazard function of the failure distribution. The decision variable of the model is the maintenance interval t for all the three maintenance policies.

3.2. Model objectives and equations

For the age policy, the integration of the reliability function R(t) over the interval [0, t] leads to the estimated expected cycle length and the up-time duration. Jardine [24] showed that the expected cycle length is equal to (the length of the preventive cycle times the probability of preventive cycle) plus (expected length of a failure cycle times the probability of a failure cycle). Sherwin [9] illustrated that when a failure occurs, it happens before the end of the preventive interval t. We need the conditional mean of the distribution given that event. Knowing that the mean value of any continuous function is by definition the first moment of area of the function relative to the origin and over appropriate intervals. This conditional mean combines the expected value and the probability of that event, as illustrated in equation (1):

In most cases, the expected cycle length

$$= t.R(t) + \int_0^t u.f(u)\mathrm{d}u\{= \int_0^t R(u)\mathrm{d}u \text{ for most distribution forms}\} \quad (1)$$

The unit cost function is defined as the expected total cost over the expected cycle length and the availability is defined as the up-time duration over the total cycle length. The model objectives used in this study are:

- For age policy:

$$\min C^{Age}(t) = \frac{C_{f1} \times F(t) \times P(t) + C_{f2} \times F(t) \times (1 - P(t)) + C_{p1} \times R(t) \times P(t) + C_{p2} \times R(t) \times (1 - P(t))}{\int_0^t R(t)}, \quad (2)$$

$$\max A^{Age}(t) = \frac{\int_0^t R(t)}{\int_0^t R(t) + T_{f1} \times F(t) \times P(t) + T_{f2} \times F(t) \times (1 - P(t)) + T_{p1} \times R(t) \times P(t) + T_{p2} \times R(t) \times (1 - P(t))}, \quad (3)$$

- For block-GAN policy:

$$\min C^{GAN}(t) = \frac{C_{f1} \times N(t) \times P(t) + C_{f2} \times N(t) \times (1 - P(t)) + C_{p1} \times P(t) + C_{p2} \times (1 - P(t))}{t}, \quad (4)$$

$$\max A^{GAN}(t) = \frac{t - \left(T_{r1} \times N(t) \times P(t) + T_{r2} \times N(t) \times (1 - P(t)) + T_{p1} \times P(t) + T_{p2} \times (1 - P(t))\right)}{t}, \quad (5)$$

- For block-BAO policy:

$$\min C^{BAO}(t) = \frac{C_{r1} \times H(t) \times P(t) + C_{r2} \times H(t) \times (1 - P(t)) + C_{p1} \times P(t) + C_{p2} \times (1 - P(t))}{t}, \quad (6)$$

$$\max A^{BAO}(t) = \frac{t - \left(T_{r1} \times H(t) \times P(t) + T_{r2} \times H(t) \times (1 - P(t)) + T_{p1} \times P(t) + T_{p2} \times (1 - P(t))\right)}{t}, \quad (7)$$

where $C^{Age}(t)$ and $A^{Age}(t)$ represent the unit cost and availability equations for age policy, $C^{GAN}(t)$ and $A^{GAN}(t)$ represent the unit cost and availability for block-GAN and $C^{BAO}(t)$ and $A^{BAO}(t)$ represent the unit cost function and availability function for block-BAO model. It is worth noting that setting $P(t) = 0$ converts the model to the traditional one, which does not consider the availability of spare parts. Thus, this model is a generalized form of the traditional one.

3.3. Solving the bi-objective function

Many methods in the literature are available for solving a set of multiple objective. In this work, we apply the weighted comprehensive criterion method [25,26], a scalarizaion technique which combines the set of bi-objective functions into one single objective function in order to minimize the deviation from the optimal value of each objective as defined in the below equations. The weighted comprehensive criterion method is adopted due to its simplicity and to the fact that a set of Pareto optimal solutions can be hereby obtained. Different optimal solutions can be obtained by controlling the importance weights τ_1 and τ_2, where $\tau_1 + \tau_2$ must equal to 1.0. Pareto diagram can be obtained by changing the values of τ_1 and τ_2.

$$\min E^{Age} = \tau_1 \times \frac{C^{Age}(t) - C^{Age}_{opt}}{C^{Age}_{opt}} + \tau_2 \times \frac{A^{Age}_{opt} - A^{Age}(t)}{A^{Age}_{opt}}, \tag{8}$$

$$\min E^{GAN} = \tau_1 \times \frac{C^{GAN}(t) - C^{GAN}_{opt}}{C^{GAN}_{opt}} + \tau_2 \times \frac{A^{GAN}_{opt} - A^{GAN}(t)}{A^{Age}_{opt}}, \tag{9}$$

$$\min E^{BAO} = \tau_1 \times \frac{C^{BAO}(t) - C^{BAO}_{opt}}{C^{BAO}_{opt}} + \tau_2 \times \frac{A^{BAO}_{opt} - A^{BAO}(t)}{A^{BAO}_{opt}}, \tag{10}$$

C_{opt} and A_{opt} represent the ultimate optimal value for the cost objective and the availability objective when solved separately,.

4. Numerical example

To test the suggested model, we assume to have a component with a lifetime described by a Weibull distribution with following characteristics: a shape parameter ($\beta = 1.6$), a scale parameter ($\eta = 10500$), $F(t) = 1 - e^{-\left(\frac{t}{\eta}\right)^{\beta}}$, $R(t) = 1 - F(t)$ and $H(t) = \int_0^t \frac{\beta}{\eta^{\beta}} t^{\beta-1}$. The calculation of $N(t)$ is done based on the modified version of the mean value theorem for integral (modified-MeVTI) proposed by Sasongko and Mahatma [27]. The probability of having spare parts is assumed to be described by an exponential distribution as in [13]. $P(t) = 1 - e^{wv}$, where w represents the work status calibration factor and v represents the ratio of the spare parts' number to the number of planned maintenance [13], v is set to 1.0. The costs and times parameters needed for the model are listed below.

$c_{f1} = 70{,}000$ $c_{p1} = 15{,}000$ $c_{r1} = 55{,}000$ $T_{f1} = 12$ $T_{p1} = 2$ $T_{r1} = 12$

$c_{f2} = 105{,}000$ $c_{p2} = 30{,}000$ $c_{r2} = 75{,}000$ $T_{f2} = 18$ $T_{p2} = 8$ $T_{r2} = 18$

In this numerical example, we present four different cases. In the first case, we consider w = 1 in the spare parts availability function. In the second case, we consider w = 2 which means lighter operation conditions. In the third case, we consider w = 1 again, but c_{p1} = 20000 instead of 15000, which results in lower C_{f1}/C_{p1}. Finally, in the last case, we consider $P(t) = 0$, i.e. we assume no spare parts to be available. Both objectives are considered equally important in all the three numerical examples.

Table 1 provides the optimal solution for solving the bi-objective model in the four different cases. When considering the cost, for all the cases, block-BAO is the most cost effective policy, followed by age, and Block-GAN. However, when considering the availability measure, the first preferred policy is age, followed by block-GAN and block-BAO. This order was obtained in the other cases, too, and can be justified by taking into account that block-BAO is the most efficient, i.e. least expensive, solution but also the less effective one with respect to the number of failures and consequently system breakdowns. For case 2, the availability results are the highest of all cases because $P(t) = 86.4\%$ for case 2, while for case 1 $P(t) = 63.2\%$. In other words, the spare parts' availability leads to higher availability result. Compared to case 1, in case 3 the ratio of C_{f1}/ C_{p1} decreased from 4.67 to 3.5, which resulted in larger replacement periods. Finally, case 4 shows the effect of spare parts' unavailability, thus, the lowest availability results are obtained with the three maintenance policies.

Table 1. Bi-objective optimal results

Case 1: w = 1 , c_{p1} = 15,000, $P(t)$ = 63.2%				**Case 3:** w=1 c_{p1} = 20,000, $P(t)$ = 63.2%			
	Age	**Block-GAN**	**Block-BAO**		**Age**	**Block-GAN**	**Block-BAO**
t	7730	8855	7212	t	8869	11524	7887
$C(t)$	7.9056	8.4050	7.5855	$C(t)$	8.1512	8.722	8.00414
$A(t)$	99.8587	99.8481	99.8337	$A(t)$	99.8594%	99.8492%	99.8327
$E(t)$	4.14×10^{-6}	5.71×10^{-6}	1.07×10^{-6}	$E(t)$	1.66×10^{-7}	3.39×10^{-7}	5.90×10^{-6}
Case 2: w = 2 , c_{p1} = 15,000, $P(t)$ = 86.4%				Case 4: w1 , and without consideration of spare parts, $P(t) = 0$			
	Age	**Block-GAN**	**Block-BAO**		**Age**	**Block-GAN**	**Block-BAO**
t	7166	7941	6737	t	8877	11554	8151
$C(t)$	6.9924	7.4037	6.7390	$C(t)$	10.3272	11.05051	9.8165
$A(t)$	99.88133%	99.87443%	99.86476%	$A(t)$	99.8037%	99.7858%	99.7546%
$E(t)$	1.39×10^{-7}	1.52×10^{-7}	6.72×10^{-6}	$E(t)$	3.92×10^{-5}	1.02×10^{-4}	1.61×10^{-6}

5. Conclusions

In this study, a bi-objective model was developed to determine the optimal replacement time for three preventive maintenance policies (age, block good as new, block bad as old) with consideration of spare parts' availability. The weighted comprehensive criterion method (WCCM) was used to deal with two conflicting objectives, namely cost and availability. The results demonstrated that the suggested model's use enables the decision maker to select the optimal maintenance policy taking into account different scenarios as well as contradicting objectives such as cost and availability, longer preventive maintenance interval, or less spare parts' inventory. As future work, it is suggested to investigate the joint optimization of production planning, inventory, and maintenance optimization and sustainability. Testing the proposed model's behavior in real applications and with other maintenance strategies such as condition monitoring could be part of the future work as well.

6. References

[1] Alsyouf I. Reliability and operation and maintenance strategies. Wind Energy Syst. Optimising *Des. Constr. Safe Reliab. Oper.*, UK: Woodhead Publishing Series in Energy; 2011, p. 303–28. doi:http://dx.doi.org/10.1533/9780857090638.3.303.

[2] Wang W. A joint spare part and maintenance inspection optimisation model using the Delay-Time concept. *Reliab Eng Syst Saf* 2011;**96**:1535–41. doi:10.1016/j.ress.2011.07.004.

[3] Wang W. A stochastic model for joint spare parts inventory and planned maintenance optimisation. *Eur J Oper Res* 2012;**216**:127–39. doi:10.1016/j.ejor.2011.07.031.

[4] Shafiee M. Maintenance strategy selection problem: an MCDM overview. *J Qual Maint Eng* 2015;**21**:378–402. doi:10.1108/JQME-09-2013-0063.

[5] Konak A, Coit D, Smith A. Multi-objective optimization using genetic algorithms: A tutorial. *Reliab Eng Syst Saf* 2006;**91**:992–1007. doi:10.1016/j.ress.2005.11.018.

[6] Alsyouf I, Hamdan S. A multi-objective optimization of maintenance policies using weighted comprehensive criterion method (WCCM). *2017 7th Int. Conf. Model. Simulation, Appl. Optim. ICMSAO 2017*, 2017. doi:10.1109/ICMSAO.2017.7934852.

[7] Barlow R, Hunter L. Optimum Preventive Maintenance Policies. *Oper Res* 1960;**8**:90–100. doi:10.1287/opre.8.1.90.

[8] Popovic VM, Vasic BM, Rakicevic BB, Vorotovic GS. Optimisation of maintenance concept choice using risk-decision factor – a case study. *Int J Syst Sci* 2012;**43**:1913–26. doi:10.1080/00207721.2011.563868.

[9] Sherwin D. Age-based opportunity maintenance. *J Qual Maint Eng* 2000;**5**:221–35.

[10] Ding S-H, Kamaruddin S, Azid IA. Maintenance policy selection model - A case study in the palm oil industry. *J Manuf Technol Manag* 2014;**25**:415–35. doi:10.1108/JMTM-03-2012-0032.

[11] Vishnu CR, Regikumar V. Reliability Based Maintenance Strategy Selection in Process Plants: A Case Study. *Procedia Technol* 2016;**25**:1080–7. doi:10.1016/j.protcy.2016.08.211.

[12] Jun Z. Maintenance planning and cost effective replacement strategies. University of Alberta, 2005.

[13] Nosoohi I, Hejazi SR. A multi-objective approach to simultaneous determination of spare part numbers and preventive replacement times. *Appl Math Model* 2011;**35**:1157–66. doi:10.1016/j.apm.2010.08.003.
[14] Sarker R, Haque A. Optimization of maintenance and spare provisioning policy using simulation. *Appl Math Model* 2000;**24**:751–60. doi:10.1016/S0307-904X(00)00011-1.
[15] Huang R, Meng L, Xi L, Liu CR. Modeling and analyzing a joint optimization policy of block-replacement and spare inventory with random-leadtime. *IEEE Trans Reliab* 2008;**57**:113–24. doi:10.1109/TR.2008.916887.
[16] Yang R, Kang J. A joint optimal policy of block preventive replacement and spare part inventory. *Int J Syst Assur Eng Manag* 2017;**8**:740–6.
[17] Lapa CMF, Pereira CMNA, Frutuoso e Melo PF. Surveillance test policy optimization through genetic algorithms using non-periodic intervention frequencies and considering seasonal constraints. *Reliab Eng Syst Saf* 2003;**81**:103–9. doi:10.1016/S0951-8320(03)00085-1.
[18] Lapa CMF, Pereira CMNA, Mol ACDA. Maximization of a nuclear system availability through maintenance scheduling optimization using a genetic algorithm. *Nucl Eng Des* 2000;**196**:219–31. doi:10.1016/S0029-5493(99)00295-2.
[19] Monga A, Zuo MJ, Toogood RW. Reliability-Based Design of Systems Considering Preventive Maintenance and Minimal Repair. *Int J Reliab Qual Saf Eng* 1997;**4**:55–71. doi:10.1142/S0218539397000059.
[20] Alsyouf I. Maintenance practices in Swedish industries: Survey results. *Int J Prod Econ* 2009;**121**:212–23. doi:10.1016/j.ijpe.2009.05.005.
[21] Raju NVS. Plant Maintenance and reliability engineering. India: Cengage Learning; 2011.
[22] Kwak RY, Takakusagi A, Sohn JY, Fujii S, Park BY. Development of an optimal preventive maintenance model based on the reliability assessment for air-conditioning facilities in office buildings. *Build Environ 2004*;39:1141–56. doi:10.1016/j.buildenv.2004.01.029.
[23] Al-Najjar B, Alsyouf I. Selecting the most efficient maintenance approach using fuzzy multiple criteria decision making. *Int J Prod Econ* 2003;**84**:85–100. doi:10.1016/S0925-5273(02)00380-8.
[24] Jardine AKS. Maintenance, Replacement and Reliability. UK: Pitmann Press; 1973. doi:10.1017/CBO9781107415324.004.
[25] Hamdan S, Cheaitou A. Dynamic green supplier selection and order allocation with quantity discounts and varying supplier availability. *Comput Ind Eng* 2017;**110**:573-589. doi: 10.1016/j.cie.2017.03.028.
[26] Hamdan S, Cheaitou A. Supplier selection and order allocation with green criteria: An MCDM and multi-objective optimization approach. *Comput Oper Res* 2017;**81**:282–304. doi:10.1016/j.cor.2016.11.005.
[27] Sasongko LR, Mahatma T. The Estimation of Renewal Functions Using the Mean Value Theorem for Integrals (MeVTI) Method. *Cartes 2016*;**5**:111–20.

Acknowledgements
The author gratefully acknowledges the help of Eng. Sadeque Hamdan from Sustainable Engineering Asset Management (SEAM) research group, and the support of the University of Sharjah, Sharjah, United Arab Emirates. In particular, the fund received from the Research Institute of Science and Engineering (RISE) for project number (1602040538-P).

IOP Conf. Series: Materials Science and Engineering **351** (2018) 012015 doi:10.1088/1757-899X/351/1/012015

Cycling transport safety quantification

Jiri Drbohlav[1] and Josef Kocourek[2]

[1] Faculty of Civil Engineering, Czech technical university in Prague, Prague, Czech Republic
[2] Faculty of Transportation Sciences, Czech technical university in Prague, Prague, Czech Republic

drbohlav.jirka@seznam.cz; kocoujos@fd.cvut.cz

Abstract. Dynamic interest in cycling transport brings the necessity to design safety cycling infrastructure. In las few years, couple of norms with safety elements have been designed and suggested for the cycling infrastructure. But these were not fully examined. The main parameter of suitable and fully functional transport infrastructure is the evaluation of its safety. Common evaluation of transport infrastructure safety is based on accident statistics. These statistics are suitable for motor vehicle transport but unsuitable for the cycling transport. Cycling infrastructure evaluation of safety is suitable for the traffic conflicts monitoring. The results of this method are fast, based on real traffic situations and can be applied on any traffic situations.

1. Safety of cycling transport

The trend in the last years shows clearly the growth of cycling transport popularity.

People discover the advantages and uniqueness of this kind of transport again.

The return to the cycling transport is realized because of several reasons. The main reason is the users are aware of the advantage of bicycle as a mean of transport within the public transport.

The increase of the bicycle users as a mean of transport brings the necessity of building a cycling infrastructure. The valid legislative and the development of integrational elements for cyclists may not adequately react to the tempo of cycling transport development. The integrational elements for cyclists in the main transport flow are the most frequent ways, how to support the bicycle users.

The development of integrational elements for the cyclists is the long-term issue which brings many difficulties. Very often, the solution that appears as suitable at the first sight may bring the decrease of cyclists safety later on. Certain element can work under specific circumstances only and it cannot be automatically and without consideration brought to other projects.

The main target of cycling transport infrastructure suggestions is the safety. The safety is the relative term, not easy to grasp. Every participant has his subjective sense of safety within the public transport and its ways. The subjective feeling influences the decision of users, when deciding which mean of transport to use and also the behaviour in it.

The objective view on the public transport safety is throughout the statistical data. These statistical data of accident rate have its clear values and are able to evaluate the safety, too.

The evaluation of cycling transport safety with the help of statistical data about the accident rate has its deficiencies. The deficiencies can lead to the completely wrong interpretation of these data.

The common statistics about the accident rate does not consider the size and quality of constructed cycling infrastructure, which is, according to the safety, crucial. Furthermore, the share of the transport labour division on the cycling infrastructure, is not considered. We call it a modal split.
According to this, it is essential, to create the safety evaluation methodology for the cycling transport, which will consider its web, the cyclists share on the total transport labour and which will evaluate the safety in real time.

2. Safety in numbers hypothesis

Safety in numbers is a hypothesis based on the idea, that if the individual is part of a big group, it is less probable, that he/she will be part of an accident, will be killed or will die due to the heart attack. Some relevant theories bring the argument, that if the individual is part of some big group, his/her behaviour is more predictable and there are less accidents.
Safety numbers hypothesis can be well applied on the cycling transport safety. Long-term researches show, that the bigger the transport labour share on the total share of transport is, the lower is the accidental rate of cyclists. It is remarkable, that the increased interest in the cycling transport "takes out" the users from the other transport types. The number of potential conflicts with motorised as well as pedestrian transport decreases.
Last but not least important fact (considering the safety of cycling transport) is that with the increased interested in cycling transport, the investments expended on building of cycling infrastructure and cycling transport itself, grow as well.

2.1. Accident rate of cycling transport in selected EU countries

Safety in Numbers methodology was used to determine accident rate of cyclists, with the modal split consideration in selected EU countries.
In the Table 1, there is a number of cyclists killed in the given year, according to selected EU countries and linked to the modal split. It is obvious, that the bigger the share of cycling transport on the total labour of transport is, the lower is the accident rate of cyclists. According to the following data, we can see that the accident rate in Netherlands is approximately 6 times lower than in United Kingdom.

Table 1. The number of cyclists killed in year per 1 million inhabitants in EU countries with a link to modal split of cycling transport, converted to 1 percent modal split [1]

Country	The number of cyclists killed in year per 1 million inhabitants a link to modal split of cycling transport			
	Modal split	2007	2008	2009
Netherlands	27	0,3	0,3	0,3
Denmark	18	0,5	0,5	0,3
Finland	11	0,4	0,3	0,3
Sweden	10	0,4	0,3	0,2
Germany	10	0,5	0,6	0,6
Belgium	8	1,0	1,0	1,0
Switzerland	6	-	0,6	1,1
Austria	5	0,9	1,5	0,9
France	3	0,7	0,8	0,8
Italy	3	2,0	1,6	1,7
Ireland	2	1,7	0,8	1,4
UK	1	2,2	1,9	1,7

IOP Conf. Series: Materials Science and Engineering **351** (2018) 012015 doi:10.1088/1757-899X/351/1/012015

The Table 2 shows the number of cyclists killed per one million inhabitants. In 2009, in Netherlands, there were 8.4 killed cyclists per 1 million inhabitants. Technically the same amount of killed per 1 million inhabitants was in Belgium 2009. The difference is in the fact, that in Netherland there is a 3 times bigger share of cycling transport on the total labour share of transport.

Table 2. The number of cyclists killed in year per 1 million inhabitants in EU countries [2]

Country	The number of cyclists killed in year per 1 million inhabitants			
	Modal split	2007	2008	2009
Netherlands	27	8,9	8,8	8,4
Denmark	18	9,8	9,8	4,5
Finland	11	4,1	3,4	3,8
Sweden	10	3,6	3,2	2,2
Germany	10	5,2	5,6	5,6
Belgium	8	8,4	8,0	8,3
Switzerland	6	-	3,4	6,7
Austria	5	4,4	7,4	4,7
France	3	2,2	2,3	2,5
Italy	3	5,9	4,8	5,0
Ireland	2	3,3	1,5	2,9
UK	1	2,2	1,9	1,7

Fig. 1 shows the distribution of the number of killed cyclists according to number of inhabitants and modal split of cycling transport. It is clear, that the bigger the share of cycling transport on the total transport labour share is, the lower the accident rate.

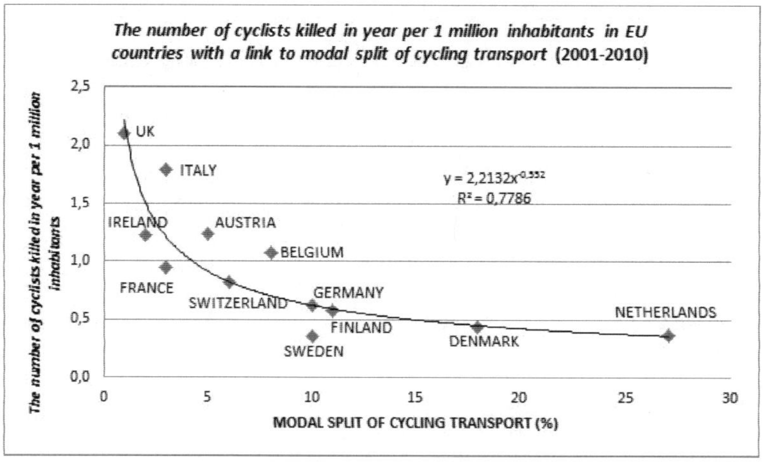

Figure 1. The number of cyclists killed in year per 1 million inhabitants in EU countries with a link to modal split of cycling transport

2.2. Accidental rate of cycling transport in the Czech republic in comparison with the other EU countries

For the methodology proposal of the cycling safety evaluation, which is being prepared in the Czech rep., is necessary to determinate current safety rates. This will be the initial for the evaluation itself. It is necessary to determine the quantification value for the evaluation of individual types (kinds) of cycling infrastructure or individual sections of communication.
To bluntly identify the position of the Czech rep. in the Fig. 1 is not easy because of the missing data about the share of cycling transport on the total transport labour. Providing that Czech rep. will have similar statistics as for example UK, Austria or France.
There were three basic premises to determinate the position of Czech rep. in Fig. 2 of cycling safety with the help of Safety in Numbers hypothesis: the share of cycling transport on the total transport labour is 2 %, 3 % or 4 %. We can gain an approximate image, what is the position of the Czech rep. (according to cycling safety) in the EU is. The logical result would be the position with the high value of accident rates.

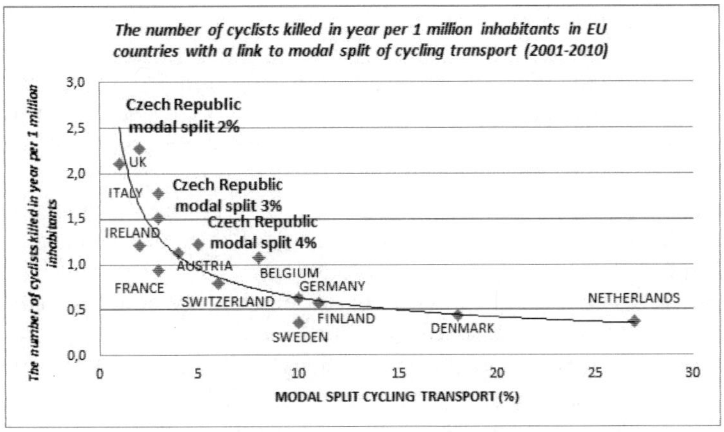

Figure 2. Anticipated ranking of the Czech rep. in the frame of Safety in numbers evaluation, according to the various shares of total transport labour.

Fig. 3 shows number of killed cyclists per 100 million kilometer. For further calculations is used number of killed cyclists per number of inhabitants.

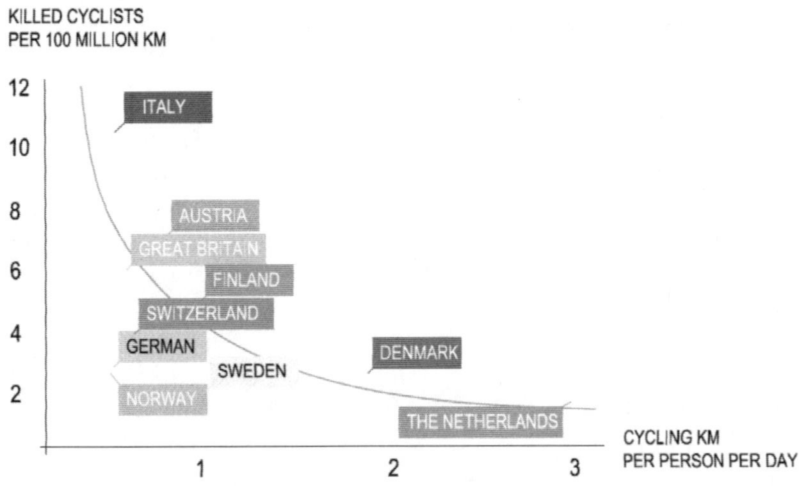

Figure 3. Killed cyclists per 100 million kilometer [1]

3. Evaluation of cycling transport safety

Evaluation of cycling transport safety with the help of monitoring methodology and evaluation of transport conflicts has unquestionable advantage in gaining the results very quickly (it is not necessary to wait for the accidents). The next unquestionable advantage is ability to monitor the real situations (conflicts) and their evaluation in real time. This can easily ang quickly affectivity of newly created integrational elements, designed for the cyclists, to increase their safety.

The next unquestionable advantage is its punctuality and the range of monitored conflicts. The statistics of accidents record only the accidents, according to the legislative of the given country. Many accidents do not appear in the statistics at all. There are relatively few accidents and thus they do not have high statistical value. What more, the statistical data react with the years delay until the data are collected to the statistical file.

With the monitoring of transport conflicts ("almost accidents") it is easy to gain much better and bigger statistical file in a split of time in comparison with current accident statistics. Thanks to this method, the cause is solved, not the result. This can be applied directly to the suggested transport safety infrastructure.

3.1. Methodology of monitoring and transport conflicts evaluation

Monitoring methodology and evaluation of transport conflict is the basic output of project "Conflict", project of public competition "Support of applied research and experimental development of ALFA", released by Technological agency of Czech republic [3].

Traditional attitude for the safety on way evaluation is monitoring of quantitative and qualitative characteristics of individual accidents. This attitude is very demanding from the financial and time-consuming point of view.

The less method used in the Czech republics is the monitoring of transport conflicts. This method monitors and evaluates individual conflict situations in given transport section.

The conflict situation is the situation very close to the accident (almost accident, but is diverted in time by i.e. changing of speed, stopping, changing the direction and so on.

The advantage is that the conflict situations are almost instantly available (after the video is recorded and analysed, it is not necessary to wait for the accident itself. The next advantage is that conflict situations without an accident are more frequent than the accident itself. Thus it is possible to grasp the more dangerous situations and locations with the help of transport research.

After the measure for the safety improvement is applied it is possible to obtain the data immediately and it is possible to analyse the individual measure. Fig. 4 shows safety pyramid, which is used in Monitoring methodology and evaluation of transport.

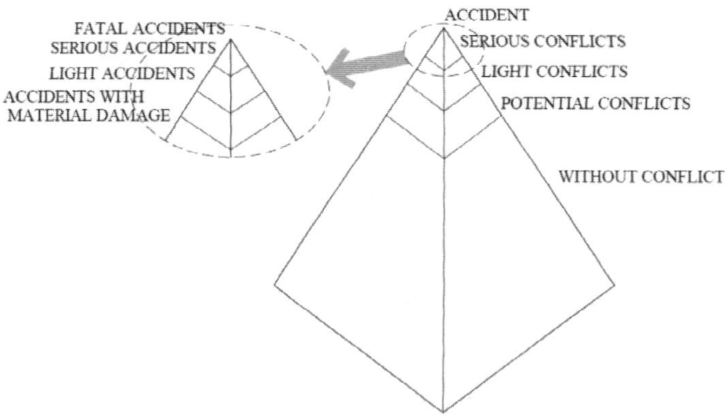

Figure 4. Safety Pyramid [4]

The conflicts are divided into five basic grades according to rules violation and jink. The conflict is defined as the situation, when the move has to changed and/or the speed of vehicle. These are two possible maneuvers to avoid the collision.
Grade "0" represents the rules violation, but with no further effects. The behaviour can have negative influence on safety (for example turning/stopping at crossing/ no respect to STOP sigh, red signal and so on). The rules are in grade "0" violated without any effect, which means there is no need of a jink or sudden speed change.
Grade "1" represents the rules violation with the reserve to handle the jink. In example it is bypassing the vehicle by the pedestrians or controlled decrease of speed of a vehicle coming to the crossing via the major road.
Grade "2" represents sudden reactions which lead to unexpected maneuvers. For example it is sudden avoidance to a vehicle.
Grade "3" represents abrupt reactions which lead to unexpected maneuvers as to try to avoid the collision in last moment.
Grade "4" represents conflict with collision (clash) result.
The seriousness is not defined by the guilt of the element which is responsible. It is the rate of the damage no matter who is guilty.

3.2. Safety in numbers hypothesis – basic safety determination

To determine the basic safety of cycling transport the Safety in Numbers hypothesis is used. On basis of the statistical file of accidents the basic rate of safety will be determined in given country. The rate of safety will be gained by the calculation of accidents followed by death.
The individual grades of accidents seriousness will be multiplied by the frequency rate coefficient according to frequency of individual accident types.
This will show the basic safety rate will be obtained. This will be initial value to be able to consider the safety, according to the monitoring of transport conflict methodology. The basic rate of safety can be determined not only for the state or country but also for the city/town or even for the given section.

4. Methodology of monitoring and evaluation of cycling transport conflicts

Methodology of monitoring and evaluation of cycling transport conflict must evaluate the behaviour of cyclists, drivers of vehicles and analyse their mutual affect.
Grade "0" represents the rules violation, but with no further effects. The rules were violated only slightly, which means the cyclists do not ride in ideal track or the car (vehicle) rides in the cycling safety lane. The next grade "1" represents the rules violation with no further effects, but the cyclist rides outside of his safety lane or dedicated location and/or the vehicle driver rides in the space dedicated for cyclists. Grade "2" represents the situation when the rule is violated. Result is the easily mastered maneuver. The next grade "3" represents the rule violation which results in sharp maneuver. The next step "4" represents the rules violation which results in sharp maneuver and avoiding the conflict in last moment. The last grade "5" represents the conflict resulting in collision.
It is expected that majority of transport conflicts will be defined as grades "0" to "2". Following assumption is, that majority of all conflicts will be defined as grade "0".
According to the recorded conflicts it is possible to obtain frequency rate/distribution of individual conflict grades. It is possible to set an assumption about how many conflicts and transitions of motor vehicles are there before the grade "5" conflict. The grade "5" conflict represents the traffic accident which ends up in material damage, health aftermaths – conflict is recorded via the Safety in Numbers hypothesis.

5. Example of using

Within the preparation of evaluation methodology of cycling transport, the transport researches were carried out. These were evaluated with the help of transport conflict methodology. As an example of cycling transport solution, two measurements gained in town "Hradec Kralove" were used.

IOP Conf. Series: Materials Science and Engineering **351** (2018) 012015 doi:10.1088/1757-899X/351/1/012015

5.1. Input data

Hradec Kralove is a very important regional city with almost 100 thousand inhabitants. Thanks to its geographical location and morphology, bike is a very popular mean of transport. Modal split is more than 16%. Hradec Kralove has a very good, nevertheless not finished, transport infrastructure network. In the city, there is a combination of elements in associated but also in the main transport space. These are frequently used. Table 3 shows number of accidents in the time period 1. 1. 2017 – 31. 10. 2015. Accidents are divided in to number of accidents, number of deaths, serious injuries and slight injuries.

Table 3. Cyclists accident frequency in the time period 1. 1. 2007 – 31. 10. 2015 in chosen cities [5]

City	Modal split	Accidents total	Deaths	Serious injuries	Slight injuries
Olomouc	6,2	391	2	18	312
Hradec Kralove	16,2	262	3	26	161
Ceske Budejovice	11,3	321	0	18	270
Pardubice	18,3	443	4	27	373
Opava	10,6	273	4	31	211
Frydek-Mistek	4,1	99	1	14	64
Prerov	14,7	150	0	27	100
Mlada Boleslav	4,2	83	0	16	55
Prostejov	22,8	167	1	17	124
Trinec	5,8	98	2	8	77

5.2. Safety in Numbers (SiN)

In Hradec Kralove has happened 262 accidents in almost 9 years. In the statistics, there are 3 accidents resulted in deaths, 26 accidents resulted in serious injuries and 161 accidents resulted in slight injuries. If we compare these data with Ceske Budejovice (city with modal split of transport cycling over 11%), we find that, in Ceske Budejovice, there are 22% more traffic accidents. Modal split is approx. 5% lower than in Hradec Kralove.

$$SiN\ (Hradec\ Kralove) = \frac{\frac{262}{94{,}242}}{16{,}2} = 0{,}17$$

$$SiN\ (Ceske\ Budejovice) = \frac{\frac{321}{93{,}883}}{11{,}3} = 0{,}30$$

$$SiN = \frac{\frac{Number\ of\ Accidents}{Number\ of\ Inhabitans}}{Modal\ Split}$$

This calculation can be specified with the help of statistical data in time period. It is necessary to consider use of the data about the share of cycling transport on the total transport labour share. The value Safety in Numbers is not fixed, it changes, depending on the size of statistical file. It is necessary to work with the same time period as the file is the same for all locations.

In Fig. 5, we can notice, that Hradec Kralove, with the result 0.17, ranks among the safest cities in given Modal Split. For approx. 16% Modal Split is the value of Safety in Number round 0.24.

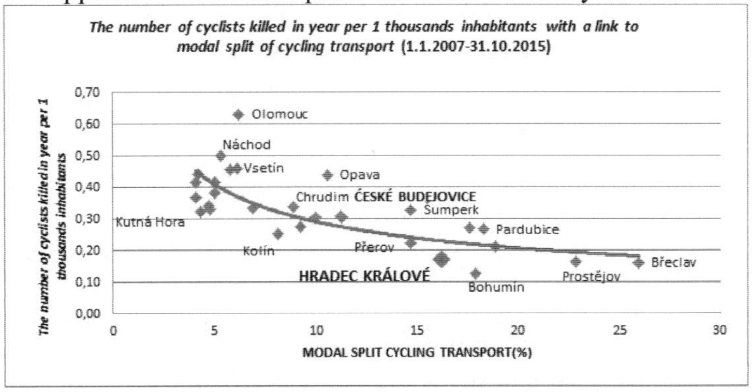

Figure 5. The number of cyclists killed in year per 1 thousands inhabitants with a link to modal split of cycling transport (1.1.2007-31.10.2015)

5.3. Monitoring methodology and traffic conflicts evaluation

Within the traffic research in Hradec Kralove, more than 11 hours in 12 sections had been recorded. In 11 hours period, there were 228 situations of mutual influence between cyclists and motor vehicles drivers. Cyclists violated the rules in 48 cases, without consequences (it means grade "0" and "1"). Motor vehicle drivers violated the rules in 19 cases in "0" and "1" grade. Out of the 228 situations, there were 67 cases of rules violations (grade "0" and "1"). During the research, 301 cyclists and 6795 motor vehicles transited the location. There were 683 situations without mutual influence between cyclists and motor vehicle drivers (grade "0" and "1"). According to the results it is evident, that 1/3 of cyclists do not ride in ideal track, or even they ride outside of reserved space. In case of motor vehicle drivers, the ratio is 1/10. These are indications, which can lead to grade "5" conflicts.

The traffic accident with the cyclist happens in Hradec Kralove once a 14 days. Modal Split considers the maximal length of cycling transport track 12km. Measured sections had recorded length of 0.1km. The total transport labour in Hradec Kralove is 23500 inhabitants, while 3800 are cyclists. Measured data represent approx.. 8% of cyclists, which participate on total transport labour share. During one day, there are 1250 cases of grade "0" and "1" caused by cyclists, with no motor vehicle interaction, 840 cases of rules violations (grade "0" and "1") caused by cyclists and motor vehicle drivers with their mutual influence.

We can expect that the grade "5" occur once in two weeks (according to statistical data). In 14 days, 17500 cases of grade "0" and "1" occur (rules violation by cyclists) and 11760 grades "1" and "0" (rules violations of cyclists and motor vehicle drives with their mutual influence). We can assume that the one traffic accident will happen out of 17500 cases, respectively 11760 according to traffic conflicts monitoring and evaluation methodology. 17500 grades (type "1" and "0") represent more than 53200 cyclists′ traffic routes. The cyclists′ traffic route is considered in Modal Split not more than 12km. Within the calculation, the average route length considered is 7km. One traffic accident of a cyclist happens after every 372400 kilometres rode. This represents the whole results for Hradec Kralove. Now, these results should be compared with the measured values on specific route profile.

5.4. Use of results

Within the bounds of methodology preparation, suitable for the cycling safety transport evaluation, individual elements of cycling infrastructure will be compared with various road profiles. Road parameters are: width of main traffic corridor, width of traffic lane and intensity of cycling, as well as motor vehicles′ infrastructure. According to these data, it will be possible to design cycling infrastructure of maximal safety.

6. Conclusion

As previously mentioned, the main and the most important essence of transport infrastructure proposal is the proposal of safety transport. Transport infrastructure is suggested according to norms, methodology and experience. With its specifications, cycling transport exceeds the common idea of transport infrastructure designs. It is due to its fast development and especially due to new elements for cyclists, growing share on the total transport labour and due to mutual influence of cyclists and motor vehicle drivers.

Cyclists´ safety evaluation with the help of traffic conflicts methodology can predict the accident rate which results in injury or death in real time. It is also able to analyse the safety of newly designed elements for cyclists and is also able to check their efficiency in a very short period of time.

On the grounds of accident rate and its data inputs, it is possible to design and suggest the most suitable and safest elements for cyclists, already in the stage of project preparation.

Within the creation of traffic conflicts methodology, the local sections of infrastructure had been observed. These were of various width of main transport corridor, intensity of motor vehicles and type of cycling infrastructure.

Methodology will be published in year 2018 within dissertation work, written by Jiri Drbohlav, student of postgraduate programme on Czech Technical University in Prague, Faculty of Civil Engineering.

7. References

[1] Fietsberaad, Cycling in the Netherlands, Ministrie van Verkeer en Waterstaat, Fietsberaad, 2009
[2] Dacota, Annual statistical report 2012, European road safety observatory, Brandstaetter, C, 2012
[3] The final research report is a partial objective 2, a study on the real share of bicycle traffic in the overall division of freight, analyza potreb budovani cyklisticke infrastruktury in Czech Republic „Cycle21“, VaV – 1F43E/045/210, MD č.: 1F43E/045/210
[4] Methodology of monitoring and evaluation of cycling transport conflict, Centrum dopravniho vyzkumu, v.v.i., Czech technical university in Prague, ISBN 978-80-86502-62-5.
[5] Web application Single vector map www.jdvm.cz

IOP Conf. Series: Materials Science and Engineering **351** (2018) 012016 doi:10.1088/1757-899X/351/1/012016

Life-Cycle environmental impact assessment of mineral industries

Shahjadi Hisan FARJANA[1], Nazmul HUDA[1,*], M. A. Parvez MAHMUD[1]

[1]Sustainable Energy Systems Engineering Group, School of Engineering, Macquarie University, Sydney, NSW-2109, Australia

nazmul.huda@mq.edu.au

Abstract. Mining is the extraction and processing of valuable ferro and non-ferro metals and minerals to be further used in manufacturing industries. Valuable metals and minerals are extracted from the geological deposits and ores deep in the surface through complex manufacturing technologies. The extraction and processing of mining industries involve particle emission to air or water, toxicity to the environment, contamination of water resources, ozone layer depletion and most importantly decay of human health. Despite all these negative impacts towards sustainability, mining industries are working throughout the world to facilitate the employment sector, economy and technological growth. The five most important miners in the world are South Africa, Russia, Australia, Ukraine, Guinea. The mining industries contributes to their GDP significantly. However, the most important issue is making the mining world sustainable thus reducing the emissions. To address the environmental impacts caused by the mining sectors, this paper is going to analyse the environmental impacts caused by the 5 major minerals extraction processes, which are bauxite, ilmenite, iron ore, rutile and uranium by using the life-cycle impact assessment technologies. The analysis is done here using SimaPro software version 8.4 using ReCipe, CML and Australian indicator method.

1. Introduction

Mining is the most dominant industrial sector in many countries with precious metal reserves. These countries already attracted industries and investors with ample opportunity for world economy, investment and development. At the same time, it is also valuable in terms of environmental effects, sustainability and global warming as the deeper it goes, the negative impact to environment increases. This issue is more feasible for those countries where mining industry in the leading industrial sector like in South Africa, Russia, Australia, Ukraine and Guinea. These countries have largest mineral and mining reserves. Similarly, China, USA, Mexico these countries are contributing as one of the largest metal producers. With 2.5 trillion-dollar worth mineral reserve, South Africa holds the largest resource and production for platinum, black coal, diamond, base metal and uranium. In Russia, with 794 billion-dollar reserves, it has resources and production for cobalt, nickel, coal and iron. The third position goes for Australia, which holds the reserve worth 737 billion-dollar. It is also the world's largest producer of bauxite, nickel. Ukraine holds 510 billion-dollar worth reserve while mostly it is iron ore. Lastly, in Guinea it worth's 222 billion-dollar reserves with bauxite deposits. On the other hand, in terms of sustainability, mining industries are harmful for air, land, water, human health and biodiversity. The major reasons behind these effects are the harmful emissions from the mining sites and mining methods, mining tailings which gets mixed with air/water, extensive mechanism and machinery used in the extraction of deep mining deposits, humans living in the mining area who are exposed to harmful emissions, emitted particles which are non-decayable and works as a restraint for plant and soil growth. During mining time, the soil and plants in the deposit area should be removed which leads to deforestation and soil erosion. Mining metal particles gets mixed with soil which restrains it from plantation thus driving towards destruction to surrounding lands. Water becomes contaminated from acid mine drainage, metal contamination, processing plants, tailing ponds, waste-disposal area, active or abandoned surface haulage etc. At the similar way, biodiversity gets affected from the massive modification of pre-mined landscape and infrastructure. This paper is going to address the environmental impacts caused by the leading mineral extraction processes in a global context through life-cycle environmental impact assessment. Section 2 illustrates the leading mineral industries, their use and environmental concerns caused by them. Section 3 elaborates life-cycle environmental impact assessment core methodologies used in this research. Section 4 describes the impact analysis results in the category of human health, global warming and ecotoxicity. Section 5 discussed the conclusion and future recommendation [1-7].

2. Mineral industries and Sustainability

Five of the major mineral industries and their manufacturing technologies are chosen here in this paper for sustainability analysis. These are bauxite, ilmenite, iron ore, rutile and uranium. These minerals and their key features are briefly described here.

Bauxite: Bauxite is the principle ore of aluminium which is not a mineral but a laterite ore. Bauxite is a mixture of aluminium oxide, hydroxide, clay minerals and insoluble metals. Bauxite is crushed and purified using the Bayer process to produce aluminium. Bauxite is commonly used as abrasive made by crushing it to form powder. Sintered bauxite is also used as oil-field propellent. The whole process of aluminium production from raw bauxite requires enormous amount of energy in the form of electricity, water and resources because of relatively stable form of aluminium. During installation, bauxite mine requires to remove all the plants and soils to start mining and extraction, which eventually lead to contamination of soils by mining tailings. During smelting and processing, greenhouse gases are released [8-10].

Ilmenite: Ilmenite is the most important ore of titanium-oxide which is considered as weakly magnetic mineral sand, grey-black in colour and solid in form. The major environmental concern associated with the mineral-sands industry is the radiation hazards. Ionizing radiation is responsible for this radiation hazards. These are mostly associated with titanium-oxide mining operations. Other issues of concern are pollution of ground-water resources, mineral transport with heavy vehicles, dredging operations in fragile coastal area and clearing of forest [11, 12, 13, and 14].

Iron: Iron in its metal form extracted from the non-metallic iron ore minerals which is a mixture and magnetic in nature. World's 95% metal is produced from iron. Iron is a shiny matter that rusts in damp air. Most of the iron is used in civil engineering for making structure and in the manufacturing sector in the form of stainless steel. The main source of environmental emissions during iron mining includes the construction and operation of iron mines which leads to several oxide emissions. However, huge amount of energy is used during iron ore extraction [11, 15, and 16].

Rutile: rutile is another most commonly found form of titanium-oxide. The common forms of the rutile are to be found in sediments, indigenous and sedimentary rocks. Rutile mining and extraction operation includes the mining from the ore body which requires cleaning using chlorine, which causes environmental problems such as toxicity, thus detrimental for habitats and plants. Moreover, like ilmenite the mining process of the rutile releases radioactive elements such as radionuclides, dust, metals and rare-earth elements [17, 18 and 19].

Uranium: Uranium is a radioactive, ductile and dense metal which can be coated with uranium oxide when in contact with air. Uranium is used widely for producing nuclear energy through chain reactions, making shields in bullets or missiles or making fertilizers. Uranium can be found in environments like soil, water or rocks. Uranium is the heaviest metal naturally found in earth's crust and is used as nuclear fuel throughout the world; previously it was used as coloring agent to decorate ceramics and glass. Over the last decades, tailings and waste management in uranium mills have been a focus, and regulations are also changing for environmental hazards and radiological risks. In addition to producing radioactive wastes, these heaps of wastes possess detrimental health effects towards the surrounding community of uranium mines and mills. The process of making yellow-cakes involves highly toxic substances. The waste rock is a slurry and excessively radioactive [20, 21 and 22].

3. Life-cycle impact analysis

Lifecycle assessment or LCA is a systematic tool to analyse and calculate the environmental effects and impacts caused by manufacturing of a product, process or activity throughout its entire lifecycle from cradle to grave, cradle to gate, gate to grave or whatever. In any life cycle assessment works, there are four basic steps which are required to be accomplished. Those are: goal and scope definition, life-cycle inventory or data collection, life-cycle impact assessment or environmental effect analysis and interpretation and recommendation of the inventory and results. In the goal and scope definition stage(ISO 14040), product/process/activity definition and description are made. The system boundary is necessary to be outlined here for detailed analysis; assumptions need to be made. Life-cycle Inventory Analysis (ISO 14041) stage consequentially tries to identify the materials, resources and energy inputs, output products, wastes and emissions. Life cycle inventory datasets are there to quantify their respective amounts per unit process included in the system boundary for life-cycle assessment. Allocation based on mass, economics or casual also might be required to be mentioned depending on the process under consideration. At the third stage, life-cycle impact assessment (ISO 14042) in the analysis phase is based on different methods like ILCD, TRACI, CML, Recipe etc. The major differences among these methods are the geographic context covered by their methodology. For example, TRACI method is designed based on North American geographic context. Similarly, ReCipe and CML method is applicable for European context. This phase assesses the impacts of the process/system/ activity on human health, ecosystem, water, land or economy. In addition, the results can be further analysed and presented through normalization, grouping or weighting of the indicator result. At the last phase of life cycle assessment, interpretation of the results and recommendation (ISO 14043) phase evaluates and interprets the results from the inventory-analysis and impact-assessment phase which compares and analyses among the production stages or the product systems and estimate the impacts per production phase. Similarly, impacts on ecosystems and environment and human health are considered and the results are presented. Uncertainty analysis and sensitivity analysis could also be carried out to decide among various cases and scenarios.

In this paper, the goal of this research work is to analyse and compare the life-cycle environmental impact of bauxite, ilmenite, iron, rutile and uranium, to access their impacts on human health, ecosystems and global warming under various categories. The scope of this research work is the inventories collected from the AusLCI database for different mines in Australia. Even though they are collected from Australian life cycle assessment database, their geographic region covered is global. The lifecycle impact assessment is done using SimaPro software version 8.4. The datasets are collected from Australian System Process database (AusLCI). The analysis methods are chosen as the Australian Indicator method, Recipe method and the CML method. These methods are the most common ones used for life-cycle impact assessments related to mining and mineral-extraction activities [22-25].

4. Results and discussion

Table 1, 2 and 3 sequentially shows the life cycle impact analysis results under human health, ecotoxicity and global warming category. The table combines the various human health impact categories covered by different life cycle assessment methods. The human health category is divided into the subcategories like carcinogenic, non-carcinogenic, global warming or ozone formation. In ecotoxicity, categories are based on the groundwater resources like marine, freshwater or aquatic. Or it can be based on the soil resources. Under the global warming category, it is categorized based on human health or ecosystems.

Table 1. Lifecycle impact analysis results on human health category using various methods.

Label	Bauxite	Ilmenite, 54% titanium dioxide	Iron ore, 46% Fe	Rutile, 95% titanium dioxide	Uranium natural
Human toxicity - non-carcinogenic	0.0001	0.0119	0.0001	0.0619	10.7958
Human toxicity - carcinogenic	0.0003	0.0155	0.0002	0.0808	12.431
Human toxicity	0.001	0.0227	0.0008	0.1177	29.1899
Global warming, Human health	0.0469	1.4946	0.0275	7.7641	100
Ozone formation, Human health	0.0349	0.2953	0.0357	1.5339	100
Human carcinogenic toxicity	0.0018	0.0358	0.0013	0.1858	6.6861
Human non-carcinogenic toxicity	0.0049	0.1586	0.0031	0.8238	28.524
Water consumption, Human health	0.0025	0.0977	0.0013	0.5073	6.9415

Table 2. Lifecycle impact analysis results on ecotoxicity category using various methods.

Label	Bauxite	Ilmenite, 54% titanium dioxide	Iron ore, 46% Fe	Rutile, 95% titanium dioxide	Uranium natural
Ecotoxicity-freshwater	0.0001	0.0047	9.07E-05	0.0243	1.6626
Ecotoxicity-marine	8.82E-05	0.003	8.09E-05	0.0154	0.192
Ecotoxicity-terrestrial	0.0307	1.2927	0.0179	6.7154	47.8598
Fresh water aquatic ecotox.	0.0004	0.0218	0.0003	0.1133	6.1769
Marine aquatic ecotoxicity	0.0005	0.0299	0.0004	0.1556	6.4993
Terrestrial ecotoxicity	0.0676	3.3479	0.0374	17.3917	79.5809
Terrestrial ecotoxicity	0.0696	2.8521	0.0415	14.816	92.4882
Freshwater ecotoxicity	0.0017	0.0573	0.0011	0.2977	11.9519
Marine ecotoxicity	0.0047	0.1478	0.0032	0.7679	20.9167

Table 3. Lifecycle impact analysis results on global warming category using various methods.

Label	Bauxite, at mine	Ilmenite, 54% titanium dioxide	Iron ore, 46% Fe, at mine	Rutile, 95% titanium dioxide	Uranium natural, at mine
Global Warming	0.0468	1.4847	0.0279	7.7124	100
Global warming (GWP100a)	0.0469	1.491	0.0279	7.7457	100
Global warming, Human health	0.0469	1.4946	0.0275	7.7641	100
Global warming, Terrestrial ecosystems	0.0469	1.4946	0.0275	7.7642	100
Global warming, Freshwater ecosystems	0.0469	1.4946	0.0275	7.7642	100

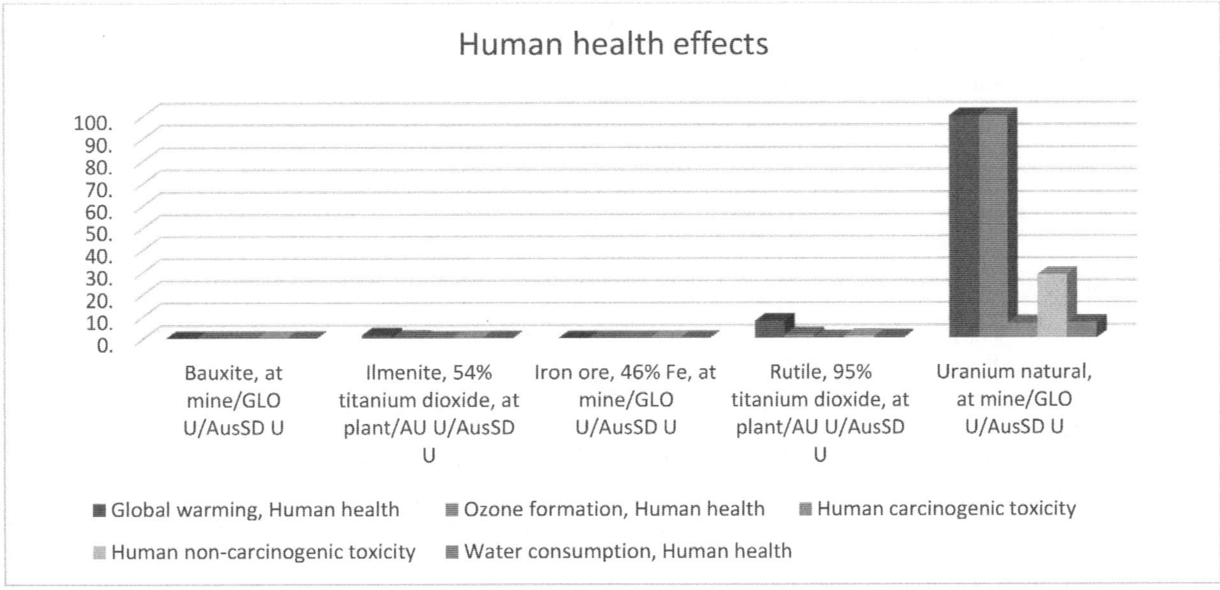

Figure 1. Human health affects results from the mineral industries.

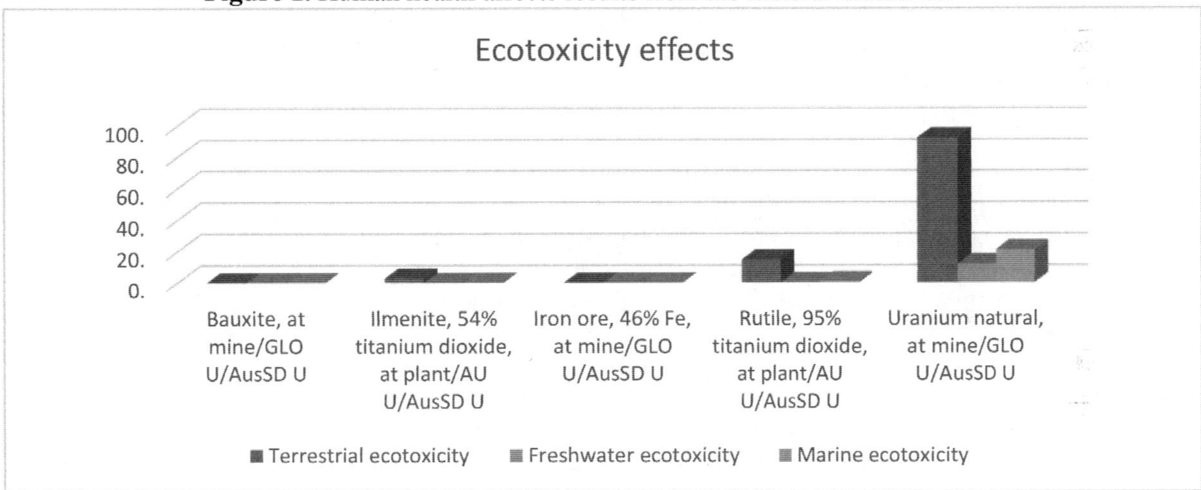

Figure 2. Ecotoxicity affects results from the mineral industries.

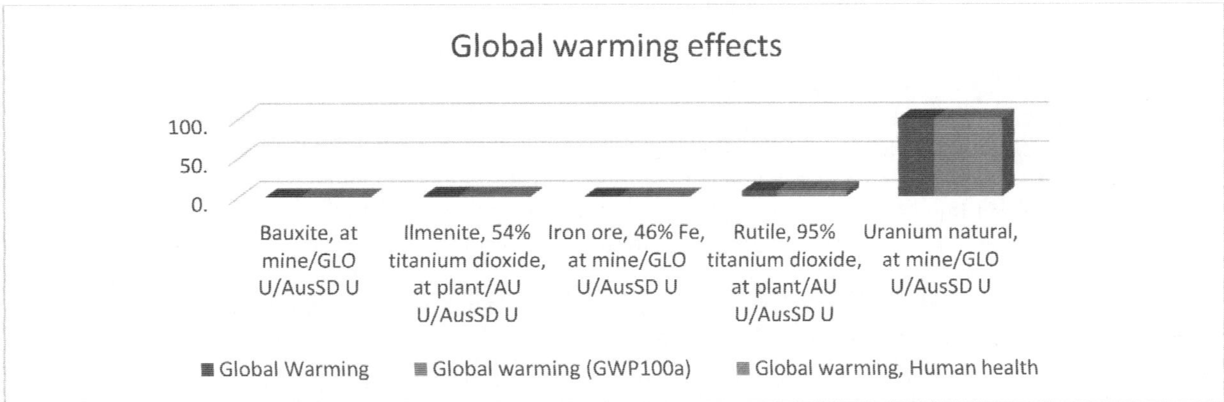

Figure 3. Global warming affects results from the mineral industries.

In all the categories of human health, ecotoxicity and global warming, uranium mining and extraction processing routes contribute mostly rather than other mineral commodities considered in this research. The reason should obviously be the radioactive particles and radionuclides emitted from the uranium processing mills and the radioactive uranium tailings. The second largest contributor is the rutile mining sector. Even though ilmenite and rutile are produced from the same mines and processing systems, rutile mining requires much more processing technologies than ilmenite mining due to ore grade and mining depth. This reason obviously draws the lesson that why rutile mining is more impactful towards environmental effects rather than ilmenite mining. Fourth position goes for bauxite and lastly comes iron ore. This trend is consistent among the considered mining industries using the CML, ReCipe and Australian indicator method. This result also indicates the consistency among the life-cycle impact assessment methods towards the mineral industries routes. These results are clearly demonstrated through Figure 1-3[22, 24, 26, 27 and 28].

5. Conclusion

In the era of global warming and sustainability, mining companies and their operating bodies should ensure their risk-free and environment-friendly mining methodologies. To quantify the impacts of mineral mining extraction and processing technologies, this paper addressed the environmental effects caused by the mineral industries and their technologies through life-cycle assessment. Using the SimaPro software and CML, ReCipe and Australian indicator methodologies, this paper indicates that the largest contributor to global warming and greenhouse gas emissions is uranium mining due to their radionuclides and radioactivity, followed by the rutile and ilmenite mining. Countries where economy is strongly dependent on mining industries, specially on minerals, should focus on to their key manufacturing technologies to protect their own environment and country from the negative effects of global warming both on their resources and human health.

6. References

[1] Abdullah N H, Mohamed N, Sulaiman L H, Zakaria T A and Abdul Rahim D 2016 Potential health impacts of bauxite mining in Kuantan *Malaysian J. Med. Sci.* **23** 1–8
[2] Abzalov M 2016 Mineral sands *Mod. Approaches Solid Earth Sci.* **12** 427–33
[3] ALS Metallurgy 2015 Mineral Sands Process Development 5
[4] Anonymus (OCED) 2010 OECD GLOBAL FORUM ON Focusing on SUSTAINABLE MATERIALS Materials Case Study 2 :Plastic
[5] Argerich J 2012 A Comparison Between a New Produced and a Remanufactured Rear Subframe *Master thesis, Uppsala Univ. Sweden.*
[6] Assessment L C, Cetim F, Neuf-brisach C and Lca G Constellium Aluminium for automotive body sheet Life Cycle Assessment summary Constellium Aluminium for automotive body sheet Life Cycle Assessment summary 1–6
[7] Awuah-Offei K and Adekpedjou A 2011 Application of life cycle assessment in the mining industry *Int. J. Life Cycle Assess.* **16** 82–9
[8] Bolowich A F 2016 Linking ecosystem services and damages from bauxite mining in an LCA context Alya Francesca Bolowich
[9] Donoghue A M, Frisch N and Olney D 2014 Bauxite Mining and Alumina Refining *J. Occup. Environ. Med.* **56** S12–7
[10] Forday G 1993 Synthetic Rutile Venture 15–7.
[11] Law E and Lane D A 1991 Boston College International and Comparative Law Review Pollution Caused by Waste from the Titanium Dioxide Industry: Directive 89/428 *C. Int'l Comp. L. Rev* **14428**
[12] Lennteach 1998 Titanium (Ti) - chemical properties, health and environmental effects
[13] Metalpedia 2017 Titanium resources, reserves and production-Metalpedia
[14] Gambogi B J 2002 By Joseph Gambogi *Europe* 1–16
[15] Haque N, Hughes A, Lim S and Vernon C 2014 Rare Earth Elements: Overview of Mining, Mineralogy, Uses, Sustainability and Environmental Impact *Resources* **3** 614–35
[16] Indian Bureau of Mines 2016 Indian Minerals Yearbook 2015 *Indian Miner. Yearb. 2015 (Part-*

III Miner. Rev. **54th Editi** 1–9
[17] Mineral Deposits Limited 2011 Corporate presentation - Zircon and Titanium mining and processing 1–25
[18] U.S. Geological Survey 2014 Titanium *Miner. Commod. Summ.* **1** 172–3
[19] USGS 2017 Titanium Mineral Concentrates *USGS Miner. Commod. Summ.* **1** 2016–7
[20] Lee A . and Nikraz H 2015 BOD: COD Ratio as an Indicator for River Pollution *Int. Proc. Chem. Biol. Environ. Eng.* **51** 139–42
[21] Mattiske A 2016 Mine rehabilitation in the Australian minerals industry Mine rehabilitation in the Australian minerals industry
[22] Mudd G M 2009 *The Sustainability of Mining in Australia: Key Production Trends and Their Environmental Implications for the Future*
[23] Navarro J and Zhao F 2014 Life-Cycle Assessment of the Production of Rare-Earth Elements for Energy Applications: A Review *Front. Energy Res.* **2** 1–17
[24] Swensen G 1996 The management of radiation hazards from the mining of mineral sands in Western Australia **17** 1–16
[25] Tesla Motors 2014 Corporate presentation *Tesla Mot. Invest. Present.* 1–33
[26] Trenton E 2012 The application of two models of life cycle assessment (LCA) for transition to the low-carbon economy: a case study in the aluminum industry
[27] International Aluminium Institute 2007 Life Cycle Assessment of Aluminium, Year 2005 Update *Update*
[28] Leigh E 2010 What Aluminum Extraction Really Does to the Environment *Electron. Recycl. Int.* 1

Trajectory data privacy protection based on differential privacy mechanism

Ke Gu[1,2], Lihao Yang[1], Yongzhi Liu[1] and Niandong Liao[1]

[1] School of Computer and Communication Engineering, Changsha University of Science and Technology, Changsha 410114, China
[2] Hunan Provincial Key Laboratory of Intelligent Processing of Big Data on Transportation, Changsha University of Science and Technology, Changsha 410114, China

gk4572@163.com

Abstract. In this paper, we propose a trajectory data privacy protection scheme based on differential privacy mechanism. In the proposed scheme, the algorithm first selects the protected points from the user's trajectory data; secondly, the algorithm forms the polygon according to the protected points and the adjacent and high frequent accessed points that are selected from the accessing point database, then the algorithm calculates the polygon centroids; finally, the noises are added to the polygon centroids by the differential privacy method, and the polygon centroids replace the protected points, and then the algorithm constructs and issues the new trajectory data. The experiments show that the running time of the proposed algorithms is fast, the privacy protection of the scheme is effective and the data usability of the scheme is higher.

1. Introduction

With the rapid development of computer and network, location data mining and analysis plays an increasingly important role in our social life. So, it is the key of developing location data services that we must be able to learn and understand position information [1]. However, trajectory (location) data is mainly collected and disseminated by mobile equipments, but many mobile devices and mobile communication technologies must integrate geographical data and individual information into trajectory data, such as individual information may contain individual privacy information, personal health status, social status and behavior habits, etc, thus mining and publishing trajectory data may divulge individual sensitive information so as to influence people's normal life. Currently many privacy protection schemes are being widely used in many fields, such as secure communication, social network, data mining and so on. The works [2, 3] first proposed the *k*-anonymity model to protect social network, whose anonymity protection methods mainly include generalization [4, 5], compression, decomposition [6], replacement [7] and interference. Based on the works of [2, 3], many other *k*-anonymous protection methods [8-16] were also proposed. However, the works [17, 18, 19] proved that some anonymous protection methods cannot protect sensitive data very well. Additionally, Cristofaro et al. [20] proposed a privacy-encrypted protection scheme. Although their scheme can ensure data security, data utility is decreased. Current location data privacy protection methods [1, 21] are mainly classified to three categories: the heuristic privacy-measure methods, the probability-based privacy inference methods and the privacy information retrieval's methods. The heuristic privacy-measure methods [22, 23, 24, 25] are mainly to provide the privacy protection measure for some no-high required users, such as *k*-anonymity [22], *t*-closing [23], *m*-invariability [24] and *l*-diversity [25]. Also, although the information retrieval's privacy protection methods can achieve perfect privacy protection, there are more or less privacy information in the released data, so these methods may result in that no data can be released, and these methods have high overhead. Additionally, the probability-based privacy inference methods can protect data and achieve better data utility under certain conditions, but the effectiveness of the methods depends on original data availability. Further, the three kinds of methods are based on a unified attack model [1], which depends on certain background knowledge to protect location data. However, with the increase of background knowledge got by the attackers, these methods could not always effectively protect location data. The works [2, 3, 8-16] showed the shortages of the relationship-privacy protection methods. Ting et al. [26] analyzed a variety of privacy threat models and tried to optimize the effectiveness of the data obtained while preventing different types of reasoning attacks. Bugra et al. [27] proposed the first effective location-privacy preserving mechanism (LPPM) that enables a designer to find the optimal LPPM for a LBS (location-based service) given user's service quality constraints against an adversary implementing the optimal inference algorithm. Such LPPM is the one that maximizes the expected distortion (error) that the optimal adversary incurs in reconstructing the actual location of a user, while fulfilling the user's service-quality requirement. Presently, it is the key of protecting location data to provide a privacy protection method not sensitively to background knowledge. Based on the requirement, differential privacy protection technology can exactly satisfy it. He et al. [28] proposed a synthetic system based on GPS path, which can provide strong differential privacy protection mechanism. The proposed system gets different speed trajectory by using a hierarchical reference method to isolate the original trajectory, and then protects the speed trajectory.

2. Trajectory data privacy protection scheme

In the section, we propose a trajectory data privacy protection scheme, which employs the differential privacy method to protect the user's trajectory data. In the proposed scheme, the algorithm first selects the protected points from the user's trajectory data; secondly, the algorithm forms the polygon according to the protected points and the adjacent and high frequent accessed points that are selected from the accessing point database, then the algorithm calculates the polygon centroids; finally, the noises are added to the polygon centroids by the differential privacy method, and the polygon centroids replace the protected points, and then the algorithm constructs and issues the new trajectory data.

2.1. Processing trajectory data

The section describes how to select the related data from the trajectory data *I* and the related and historic point data set *D*. The proposed algorithm selects the protected point data $f \in A$ and its adjacent points from *D*. Figure 1 shows the procedure of selecting the related data.

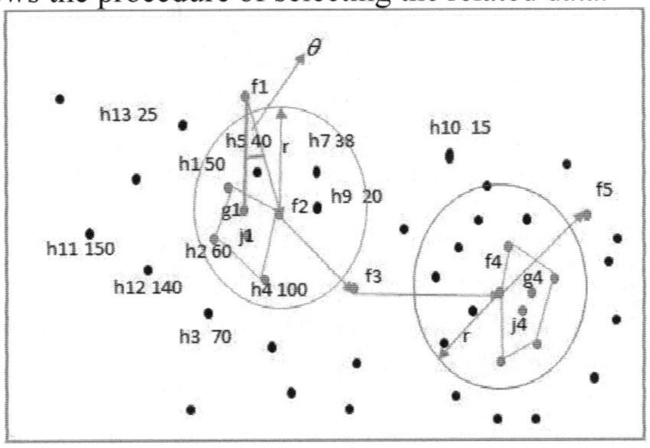

Figure 1. Processing Trajectory Data

In Figure 1, a random trajectory of one user is shown, where the points of symbol f are formed to the trajectory, and the points of symbol h denote the accessed location points (where the adjacent points must not be accessed by the current user), which build the related and historic point data set D (the historic duration is in one month). According to the Figure 1, the procedure of selecting the related data may be described as follows:

a) The proposed algorithm inputs the trajectory data I of one user, the related and historic point data set D and the related privacy protection parameters r, ε and *min_count*;
b) The algorithm selects the protected point set A from the trajectory data I;
c) The proposed algorithm forms the point set B according to the point data $f_i \in A$ and its corresponding adjacent points from D, where the adjacent points belong to the range of a circle that f is the center of the circle and r is the corresponding radius, and the frequent accessing count of the adjacent points is no less than *min_count*.

2.2. Building polygon model

The section describes how to build the polygon model to compute the polygon centroid. The proposed algorithm forms the polygons according to the protected points $f \in A$ and the corresponding adjacent points from D. Figure 1 also shows procedure of forming polygon. In Figure 1, the trajectory of one user is $f_1, f_2, \ldots\ldots f_5 \in I$, and the points $h_1, h_2, \ldots\ldots h_{13}$ with accessing counts come from D, where $f_2, f_4 \in A$ are the protected points. In the green circle that f_2 is the center of the circle and r is the corresponding radius, the points h_1, h_2 and h_4 ($\in D$ and their accessing counts >=50) and the point f_2 are formed to a polygon. Then the proposed algorithm computes the polygon centroid j_1 (noises are added to j_1 to generate a new point g_1). In fact, the algorithm may traverse the set B to build the polygons. We need to remark that the points h_1, h_2 and h_4 is nearby the point f_2, thus the points be used to build the polygon so as to maintain the usability of the modified trajectory, and that *min_count* are 50, thus some points whose accessing counts are less than 50 are not used to form the polygon in the green circle, such may distort the attacker's goal and improve the efficiency of the proposed scheme. In this paper, we set that $P_i(x_k, y_k)$ is the coordinate of the k-th vertices of the i-th polygon, $|P_i|$ is the vertices number of the i-th polygon, and $j_i(x, y)$ is the coordinate of the i-th polygon centroid. The polygon centroids are built to the set J, where $j_i(x, y) \in J$.

2.3. Adding noises based on the laplace's mechanism

In the section, we show how to use the Laplace's mechanism to add the noise $Lap((k \cdot \Delta Q)/\varepsilon)$ into the set J, where ΔQ is the sensitivity of the query function Q, where we set $\Delta Q = \text{MAX}\{\sqrt{(P_i.x_k - j_i.x)^2 + (P_i.y_k - j_i.y)^2}\}$ with $i=1,2,\ldots\ldots|N_P|$ and $k=1,2,\ldots\ldots|P_i|$, the $|N_P|$ is the number of the polygons and the $|P_i|$ is the number of the vertices of every polygon. The main steps of the algorithm are described as follows:

a) Input the privacy protection level ε and the polygon centroid set J, and then generate the noise $Lap((k \cdot \Delta Q)/\varepsilon)$ satisfying the probability $\Pr(j(x, y), \lambda)$, where $\Pr\left(j(x,y),\lambda\right) = \frac{1}{2\cdot\lambda}\cdot e^{\frac{-|j(x,y)|}{\lambda}}$. In the above formula, the variant $j(x,y)$ denotes the corresponding coordinate of the polygon centroid and $\lambda = (k \cdot \Delta Q)/\varepsilon$.
b) Add the noise $Lap((k \cdot \Delta Q)/\varepsilon)$ into the set J so as to disturb the polygon centroids. We need to remark if the formed polygon is on the left of the protected point from the trajectory data I, then the operation "+" is used; otherwise, the formed polygon is on the right of the protected point from the trajectory data I, then the operation "-" is used:

$$j_i.x = j_i.x \pm Lap\left((k \cdot \Delta Q)/\ \varepsilon\right),$$
$$j_i.y = j_i.y \pm Lap\left((k \cdot \Delta Q)/\ \varepsilon\right),$$

IOP Conf. Series: Materials Science and Engineering **351** (2018) 012017 doi:10.1088/1757-899X/351/1/012017

where $j_i \in J$, $j_i(x, y)$ denotes the coordinate of the i-th polygon centroid, and $Lap((k \cdot \Delta Q)/\varepsilon)$ is each round of the independent noise subjecting to the probability $\Pr(j(x, y),\lambda)$. Finally, the algorithm generates the set G.

Use the modified polygon centroids from G to replace the correspondingly protected points $f \in A$, and then issue the new trajectory data I'. For example, as Figure 2 shown, noise is added to j_1 to generate a new point g_1, and then g_1 is used to replace the point f_2, thus the original trajectory $f_1 \rightarrow f_2 \rightarrow f_3$ changes to $f_1 \rightarrow g_1 \rightarrow f_3$.

3. Experiment and efficiency analysis

3.1. Running time analysis

In the section, we test the running time of the proposed algorithms mainly through the time of extracting the available data, namely we mainly test the effectiveness of computing all the polygon centroids from the available data. In the tests, when we set r=70 and ε=1, 2, 3, 4, 5, 6, 7, 8, 9 respectively, the time of extracting the available data is described as Table 1.

Table 1. The efficiency of extracting available data

ε	1	2	3	4	5	6	7	8	9
Time (ms)	4	4	3	3	3	4	3	3	3

From the Table 1, we may know the time of extracting the available data is very fast, and the efficiency of computing all the polygon centroids from the available data is always increasing with the increasing of ε in a certain range.

3.2. Protection effectiveness analysis

In the section, we test the protection effectiveness of the proposed algorithms mainly through the trajectory deviation rate and the trajectory accurate rate, where the trajectory deviation rate is the angle θ formed by the modified polygon centroid and the original trajectory points, and the trajectory deviation rate is bigger in a certain range, the protection effectiveness is higher; the trajectory accurate rate is to test the protection effectiveness and usability of the noise-added data, the trajectory accurate rate is smaller in a certain range, the usability is higher.

In the test, we compute the trajectory accurate rate through the following methods: 1) set the coordinate (a_i, b_i) of the polygon centroid; 2) compute the hypotenuse $c_i = \sqrt{a_i^2 + b_i^2}$; 3) compute the accurate rate $Z = \left|1 - c_i'/c_i\right|$, where c_i is the original hypotenuse and c_i' is the noise-added hypotenuse. The trajectory deviation rate is bigger in a certain range, the protection effectiveness is higher; the trajectory accurate rate is smaller in a certain range, the usability is higher. So, when we fixedly set ε=5 and r=40, 50, 60, 70, 80, 90, 100, 110 respectively, Table 2 shows the deviation rate and accurate rate of the trajectory data.

Table 2. Trajectory deviation rate and accurate rate (ε=5)

r	c_i	c_i'	Z	θ
40	645.264	613.125	0.049807	23.2510
50	645.264	613.125	0.049807	23.2510
60	645.264	613.125	0.049807	23.2510
70	645.264	613.125	0.049807	23.2510
80	645.264	613.125	0.049807	23.2510
90	608.511	572.839	0.058621	24.7920
100	608.511	572.839	0.058621	24.7920
110	608.511	572.839	0.058621	24.7920

From the Table 2, we may know that the polygon centroid is not changed with the increasing of r when $r<90$, thus the deviation rate θ and the accurate rate Z are also not changed. Such shows that in the range of $r<90$, the new points are not selected to form the new polygon, thus the polygon is not modified. When $r>=90$, the new points are selected to form the new polygon, thus the polygon centroid is recomputed, thus the deviation rate θ and the accurate rate Z are changed. Such shows that the deviation rate θ could become big with the increasing of r, but the data usability becomes small.

4. Conclusions

In this paper, we propose a trajectory data privacy protection scheme based on differential privacy mechanism. The proposed algorithm first selects the protected points from the user's trajectory data; secondly, the algorithm forms the polygon according to the protected points and the adjacent and high frequent accessed points that are selected from the accessing point database, then the algorithm calculates the polygon centroids; finally, the noises are added to the polygon centroids by the differential privacy method, and the polygon centroids replace the protected points, and then the algorithm constructs and issues the new trajectory data. The experiments show that the privacy protection of the scheme is effective and the data usability of the scheme is higher.

5. Acknowledgment

This study is funded by the National Natural Science Foundations of China (No.61402055, No.61504013) and the Scientific Research Project of Hunan Provincial Education Department (No.15C0041, No.12C0010).

6. References

[1] Ouyang Jia, Yin Jian, Liu Shaopeng, Liu Yuba. An Effective Differential Privacy Transaction Data Publication Strategy. *Journal of Computer Research and development*, 2014, 51(10):2195-2205.

[2] Samarati P, Sweeney L. Generalizing data to provide anonymity when disclosing information. *Proc. of the 7th ACM SIGACTSIGMOD-SIGART Sympon Principles of Database Systems*, 1998, 188-202.

[3] Sweeney L. K-Anonymity: A model for protecting privacy. *Int'l Journal on Uncertainty, Fuzziness and Knowledge-Based System*, 2012, 10(5):557-570.

[4] Fung BC, Wang K, Yu PS. Anonymizing classification data for privacy preservation. *IEEE Trans on Knowledge and Data Engineering*, 2007,19(5):711-725.

[5] Sweeney L. Achieving k-anonymity privacy protection using generalization and suppression. *International Journal of Uncertainty, Fuzziness and Knowledge-Based Systems*, 2012, 10(5):571-588.

[6] Xiao X, Tao Y. Anatomy. Simple and effective privacy preservation. *Proc of VLDB 2006, New York: ACM*, 2006, 139-150.

[7] Zhang Q, Koudas N, Srivastava D, et al. Aggregate query answering on anonymized tables. *Proc of the 23rd Int Conf on Data Engineering(ICDE)*, Piscataway, NJ: IEEE, 2007, 116-125.

[8] Wong RCW, Li J, Fu AWC, Wang K. (a, k)-Anonymity: An enhanced k-anonymity model for privacy-preserving data publishing. *Proc.of the ACM 12th SIGKDD Int'l Conf on Knowledge Discovery and Data Mining*, 2006, 754-759.

[9] LeFevre K, DeWitt DJ, Ramakrishnan R. Mondrian multidimensional k-anonymity. *Proc.of the 22nd Int'l Conf. on Data Engineering*, 2006, 6(3): 25-35.

[10] Machanavajjhala A, Gehrke J, Kifer D, Venkitasubramaniam M. L-Diversity: Privacy beyond k-anonymity. *Proc.of the 22nd IEEE Int'l Conf. on Data Engineering*, 2006.

[11] Xiao XK, Tao YF. Anatomy: Simple and effective privacy preservation. *Proc.of the 32nd Int'l Conf. on Very Large Data Bases*, 2006, 139-150.

[12] Xiao X, Tao Y. Personalized privacy preservation. *Proc.of the 2006 ACM SIGMOD Int'l Conf. on Management of Data*, 2006, 229-240.

[13] Xu J, Wang W, Pei J, Wang X, Shi B, Fu AWC. Utility-Based anonymization using local recoding. *Proc.of the 12th ACM SIGKDD Int'l Conf. on Knowledge Discovery and Data Mining*, 2006, 785-790.

[14] Li N, Li T, Venkatasubramanian S. t-closeness: Privacy beyond k-anonymity and l-diversity. *Proc.of the 23rd IEEE Int'l Conf. on Data Engineering*, 2007, 106-115.

[15] Wong RCW, Fu AWC, Wang K, Pei J. Minimality attack in privacy preserving data publishing. *Proc.of the 33nd Int'l Conf. on Very Large Databases*, 2007, 543-554.

[16] Tao Y, Xiao X, Li J, Zhang D. On anti-corruption privacy preserving publication. *Proc.of the 24th Int'l Conf. on Data Engineering*, 2008, 725-734.

[17] Backstrom L, Dwork C, Kleinberg J. Wherefore are thouR3579X?: Anonymized social networks, hiddern patterns and structural steganography. *Proc.of the 16th Int'l Conf. on World Wide Web*, 2007, 181-190.

[18] Zheleva E, Getoor L. Preserving the privacy of sensitive relationships in graph data. *Proc.of the 1st ACM SIGKDD Workshop on Privacy, Security, and Trust in KDD*, 2007, 153-171.

[19] Korolova A, Motwani R, Nabar SU, Xu Y. Link privacy in social networks. *Proc. of the 24th Int'l Conf. on Data Engineering*, 2008, 1355-1357.

[20] De Cristofaro E, Soriente C, Tsudik G, et al. Hummingbird: Privacy at the time of twitter. *IEEE Symposium on Security and Privacy-S&P*, 2012, 285-299.

[21] Beresford AR, Rice A, Skehin N, Sohan R. MockDroid: Trading privacy for application functionality on smartphones. *Proc. of the 12th Workshop on Mobile Computing Systems and Applications*, ACM Press, 2011, 49-54.

[22] Huo Z, Meng XF. A survey of trajectory privacy-preserving techniques. *Chinese Journal of Computers*, 2011, 34(10):1820-1830.

[23] Bamba B, Liu L, Pesti P, Wang T. Supporting anonymous location queries in mobile environments with privacy grid. *Proc. of the 17th Int'l Conf. on World Wide Web*, New York: ACM Press, 2008, 237-246.

[24] Liu L. From data privacy to location privacy: Models and algorithms. *Proc. of the 33rd Int'l Conf. on Very Large Data Bases*, New York: ACM Press, 2007, 1429-1430.

[25] Liu F, Hua KA, Cai Y. Query l-diversity in location-based services. *Proc. of the 10th Int'l Conf. on Mobile Data Management*, Taipei, 2009, 436-442.

[26] Ting W, Ling L. From Data Privacy to Location Privacy. *Machine Learning in Cyber Trust*, Berlin: Springer, 14 March 2009, pp.217-246.

[27] Bugra G, Ling L. Protecting Location Privacy. *IEEE Transactions on Mobile Computing*, 2008, 7(1):1-18.

[28] He X, Cormode G, Machanavajjhala A, Procopiuc CM, Srivastava D. Differentially Private Trajectory Synthesis Using Hierarchical Reference Systems. *VLDB Journal*, 2015, 8(11):1154-1165.

IOP Conf. Series: Materials Science and Engineering **351** (2018) 012018 doi:10.1088/1757-899X/351/1/012018

Residual translation compensations in radar target narrowband imaging based on trajectory information

Wenjue Yue, Bo Peng*, Xizhang Wei, Xiang Li and Dongping Liao

Research Institute of Space Electronics, National University of Defense Technology, Changsha, China

pengbo06@gmail.com

Abstract. High velocity translation will result in defocusing scattering centers in radar imaging. In this paper, we propose a Residual Translation Compensations (RTC) method based on target trajectory information to eliminate the translation effects in radar imaging. Translation could not be simply regarded as a uniformly accelerated motion in reality. So the prior knowledge of the target trajectory is introduced to enhance compensation precision. First we use the two-body orbit model to figure out the radial distance. Then, stepwise compensations are applied to eliminate residual propagation delay based on conjugate multiplication method. Finally, tomography is used to confirm the validity of the method. Compare with translation parameters estimation method based on the spectral peak of the conjugate multiplied signal, RTC method in this paper enjoys a better tomography result. When the Signal Noise Ratio (SNR) of the radar echo signal is 4dB, the scattering centers can also be extracted clearly.

1. Introduction

High velocity translation induces propagation delay of the echo, which subject micro-Doppler to translation, slope and leads to defocusing phenomena in radar imaging [1]. Thus, translational compensation is a crucial step in micro-Doppler extraction and radar imaging processing.

The existing researches on translation compensation can be divided into two aspects: (1) Conduct translation compensation according to the Doppler spectrum. Y. C. Yang proposes a translation parameter estimation method based on the Doppler information of the strongest scattering point [2]. S. S. He proposes a translation parameters estimation method based on the spectral peak information [3]. He designs a quadratic equation to model the translation, and then use the spectral peak position to estimate the velocity and acceleration based on the conjugated multiplied signal. X. Q. Luo proposes a compensation method for ISAR imaging based on short time adaptive Gaussian chirplet decomposition [4]. (2) Estimate the translational parameters of the target. Reference [1] and [5] have analyzed the acceleration effects on the micro-Doppler. They proposed a compensation method based on spectral rearrangement, which can only be used for uniform motion and uniformly accelerated motion. However, most of the researches are based on the assumption that the translation is simple motion, which is difficult to achieve in reality.

Focus on the translation compensation problem, we introduce a priori information of the target trajectory, and then use the two-body orbit model to enhance the compensation precision. The tomography results are used to confirm the validity of the method. The paper is organized as follows: In section II, we figure out the trajectory of target, and then build the radar echo model of the micro-motion target. In section III, the estimated trajectory information is applied to deduce the propagation delay of the high velocity translation. Then, we conduct stepwise compensations on the residual translation until it achieves the required precision. Finally, we use tomography to evaluate the compensation results. Experimental results are exhibited in Section IV to demonstrate the necessity of the proposed method. We conclude this paper in Section V.

2. Signal model

2.1. Trajectory of the target in two-body orbit model

Suggest that earth and the radar target are regarded as the two-body system, the orbit model of the target is referred to as a two-body orbit model. During its flight, only gravity is taken into consideration, the radar target is doing non-distribution motion. As is shown in Figure 1, the trajectory of the radar target is part of an ellipse [6].The trajectory can be represented by six integrated constants, which is called orbital element. The most frequently used orbital element is $\sigma = (a, e, i, \Omega, \omega, \tau)$ [6]. The orbital element solving process is as follows.

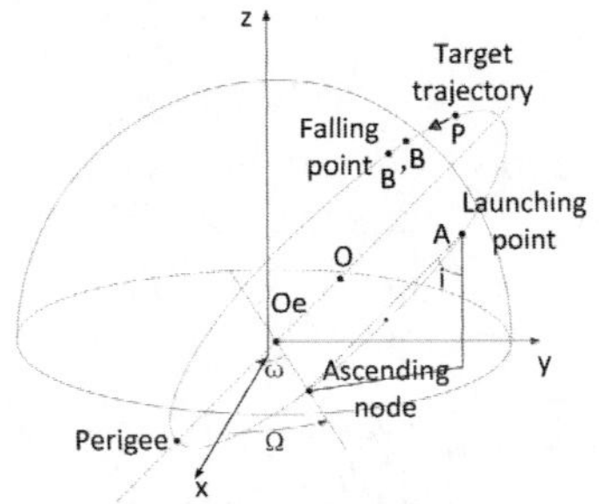

Figure 1. Geometric relation of orbit parameters.

In Earth Centered Inertial (ECI), we suggest that the position vector of point k at time t_k is $\vec{r}_k = (p_x, p_y, p_z)^T$, the velocity vector is $\vec{v}_k = (v_x, v_y, v_z)^T$, the radar target flies at the optimal angle. The geocentric distance is $r_k = \sqrt{p_x^2 + p_y^2 + p_z^2}$, the velocity is $v_k = \sqrt{v_x^2 + v_y^2 + v_z^2}$ [7]. The semi-major axis of the elliptic orbit can be written as,

$$a = (\frac{2}{r_k} - \frac{v_k^{\ 2}}{\mu})^{-1} \tag{1}$$

where $\mu = 3.9860044 \times 10^{14} m^3 / s^2$, it is the geocentric gravitational constant.
The eccentricity of the elliptic orbit can be written as,

$$e = [(\frac{r_k v_k}{\sqrt{a\mu}})^2 + (1 - \frac{r_k}{a})^2]^{1/2} \tag{2}$$

The time of passing through the perigee τ can be expressed as,

$$\tau = t_k - \frac{E_k - e \sin E_k}{n} \tag{3}$$

where $E_k = \arctan \frac{\sqrt{a} r_k v_k}{\sqrt{\mu}(a - r_k)}$ is the eccentric anomaly of point k , $n = \sqrt{\frac{\mu}{a^3}}$ is the mean angular velocity.

The right ascension of ascending node Ω can be expressed as [8],

$$\Omega = \arccos(\frac{-H_{yk}}{\sqrt{H_{xk}^{\ 2} + H_{yk}^{\ 2} + H_{zk}^{\ 2}}}) \tag{4}$$

where the moment of orbital momentum $\vec{r}_k \times \vec{v}_k = \begin{bmatrix} H_{xk} \\ H_{yk} \\ H_{zk} \end{bmatrix}$.

The included angle of orbital plane and equatorial plane i can be expressed as,

$$i = \arccos(\frac{H_{zk}}{\sqrt{H_{xk}^2 + H_{yk}^2 + H_{zk}^2}}) \tag{5}$$

The included angle of perigee and ascending node ω can be expressed as,

$$\omega = \arctan(\frac{p_z / \sin i}{p_x \cos\Omega + p_y \sin\Omega}) - f_k \tag{6}$$

where $f_k = 2\arctan\left(\sqrt{\frac{1+e}{1-e}} \tan\frac{E_k}{2}\right)$.

To summarize, $\sigma = (a, e, i, \Omega, \omega, \tau)$ is obtained.

The Kepler's equation can be expressed as,

$$M = E - e\sin E \tag{7}$$

Equation (7) is a transcendental equation of E, it is hard to figure out E accurately. So we use Newton iterative method to solve Kepler's equation,

$$E_{i+1} = E_i - \frac{E_i - e\sin E_i - M}{1 - e\cos E_i} \quad (i = 0,1,2\cdots) \tag{8}$$

where E_0 is the eccentric anomaly at time $t_k = 0$, the condition of convergence is $|E_{i+1} - E_i| \le \varepsilon$. From equation (7) and (8) we can get the relationship between E and t.

In the two-body model, the position vector of the radar target can be expressed as,

$$\vec{r}_l = \begin{bmatrix} X \\ Y \\ Z \end{bmatrix} = a(\cos E - e)\cdot\vec{P} + a\sqrt{1-e^2}\sin E \cdot \vec{Q} \tag{9}$$

where

$$\begin{cases} \vec{P} = \begin{bmatrix} \cos\Omega\cos\omega - \sin\Omega\sin\omega\cos i \\ \sin\Omega\cos\omega + \cos\Omega\sin\omega\cos i \\ \sin\omega\sin i \end{bmatrix} \\ \vec{Q} = \begin{bmatrix} -\cos\Omega\cos\omega - \sin\Omega\sin\omega\cos i \\ -\sin\Omega\cos\omega + \cos\Omega\sin\omega\cos i \\ \cos\omega\sin i \end{bmatrix} \end{cases} \tag{10}$$

We can figure out the relationship between the position vector $\vec{r}$ and t from equation (9).

$$r_{LOS}(t) = |\vec{r}_l - \vec{r}_{radar}| \tag{11}$$

where $\vec{r}_{radar}$ is the position vector of the radar.

2.2. *Radar echo signal model of the micro-motion target*

Micro-motion includes vibration, rotation, precession, nutation [9]. In Figure 2, we take precession for an example. Suggest that there is a micro-motion target in the reference coordinate system $o-xyz$, o is the origin. P_1 is the scattering center of the conic node. P_2 and P_3 are the scattering centers of the conic bottom. Precession angular velocity is ω_v, precession angle is θ_v, the included angle between radar line of sight (LOS) and the precession axis is γ, initial phase is ϕ_0. Distance between P and the radar is R_0. The aspect angle of the target is $\beta(t)$.

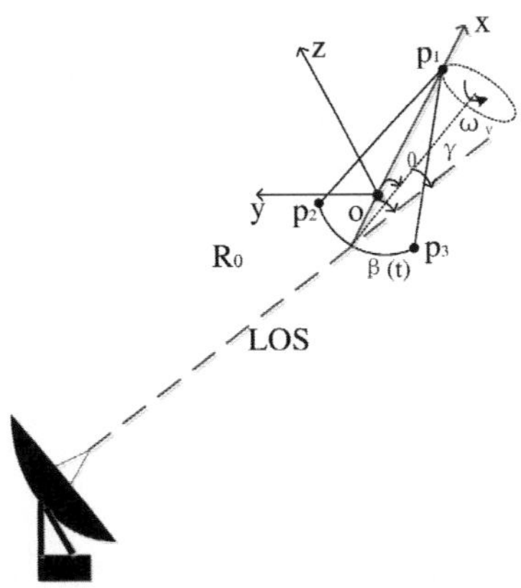

Figure 2. Micro-motion detected by radar.

$\beta(t)$ can be expressed as,

$$\beta(t) = \arccos\left(\cos\gamma\cos\theta_v + \sin\gamma\sin\theta_v\cos(\omega_v t + \phi_0)\right) \quad (12)$$

Owing to the precession, the distance changes of the target scattering center can be expressed as,

$$r_v(t) = x_m\cos\beta(t) + y_m\sin\beta(t) \quad (13)$$

where (x_m, y_m) is the position of the scattering center m.

Based on the analysis above, the radar echo signal of the target can be expressed as,

$$s(t) = \sum_{m=1}^{M}\sigma_m \exp\left(-j\frac{4\pi f_c}{c}\left(r_{LOS}(t) + r_v(t)\right)\right) + w[n] \quad (14)$$

where f_c is the radar center frequency, c is the velocity of light, σ_m is the scattering coefficient of the scattering center. $r_{LOS}(t)$ is the position change of the target caused by high velocity translation, $r_v(t)$ is the position change of the scattering center caused by micro-motion, $w[n]$ is the white Gaussian noise.

3. RTC method

In order to revise the time-frequency spectrums, we need to eliminate the effects caused by the high velocity translation. Therefore, a Residual Translation Compensations (RTC) method is proposed in this section. The steps are as follows.

- Step 1: Setup parameters of the target to figure out the trajectory and the radial distance $r_{LOS}(t)$ according to section IIA .
- Step 2: Then we use the estimated radial distance $r_{LOS}(t)$ to eliminate the effects of translation by conjugate multiplication method as follows.

$$s_{LOS}^*(t) = \exp\left(j\frac{4\pi f_c}{c}r_{LOS}(t)\right) \quad (15)$$

$$s'(t) = s(t)\cdot s_{LOS}^*(t) \quad (16)$$

where $s'(t)$ is the radar echo signal after the first time compensation by using the estimated trajectory.

- Step 3: The word "data" is plural, not singular. Use stepwise compensations method to deal with the residual translation. Suggest that the time delay before being compensated is $t_t(i)$, the time delay after being compensated is $t_e(i)$, the deviation is,

$$\begin{cases}\Delta_{i+1} = t_t(i) - t_e(i) - \Delta_i & i = 1,2,\cdots n\\ \Delta_1 = 0\end{cases} \quad (17)$$

Suggest that Δ is a quadratic equation, then we can figure out the parameters of equation.

- Step 4: If $\max\left(\left|\frac{\Delta_{i+1}}{t_t(i)}\right|\right) > 10^{-5}$, repeat step 3. If $\max\left(\left|\frac{\Delta_{i+1}}{t_t(i)}\right|\right) \le 10^{-5}$, the required precision is obtained after i times compensation. Finally, the radar echo signal can be expressed as,

$$s'_{final}(t) = s(t) \cdot s^*_{LOS}(t) \cdot \prod_{i=i}^{n} \exp(j2\pi f_c \cdot \Delta_i) \tag{18}$$

- Step 5: The time-frequency spectrum of micro-Doppler can be obtained by using STFT on $s'_{final}(t)$.

$$[mD] = STFT(s'_{final}(t)) \tag{19}$$

The flow chart is as follows (Figure 3).

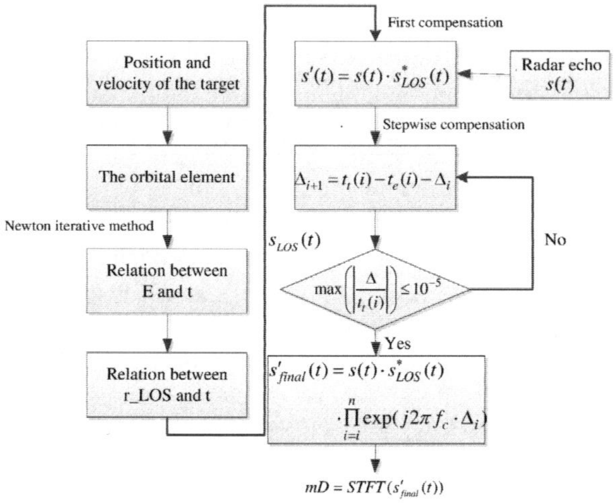

Figure 3. Flow chat of the algorithm.

4. Simulation and analysis

4.1. Simulation on trajectory of the target

When $t_k = 0$, suggest that the launching position is $(-180340.978m, 2786423.914m, 5727756.069m)$, the velocity is $(-1935.96m/s, 1300.56m/s, 4644.84m/s)$, the target flies in the optimum trajectory, it flies for 1128.4s. Radar is close to the landing position. The three-dimensional trajectory of the target in ECI is shown in Figure 4.

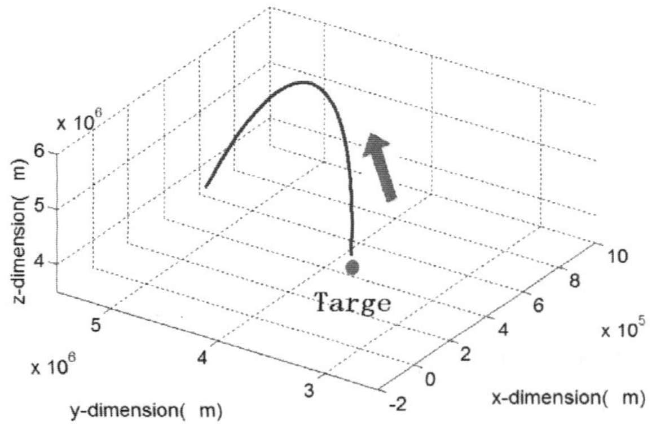

Figure 4. The three-dimensional trajectory of the target in ECI.

The changes of height and the actual velocity are shown in Figure 5. The radial distance between radar and the target is shown in Figure 6.

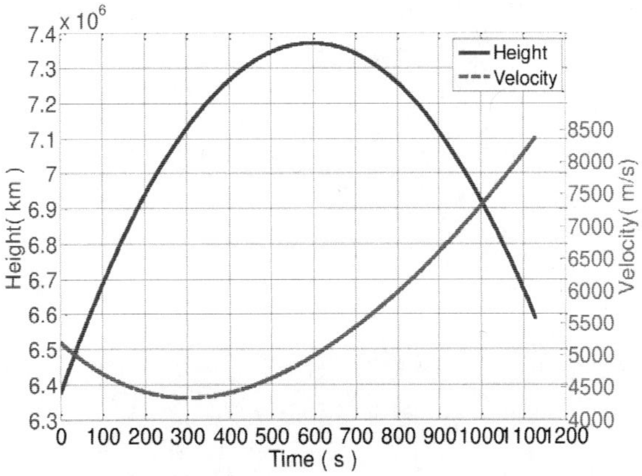

Figure 5. The changes of height and the actual velocity.

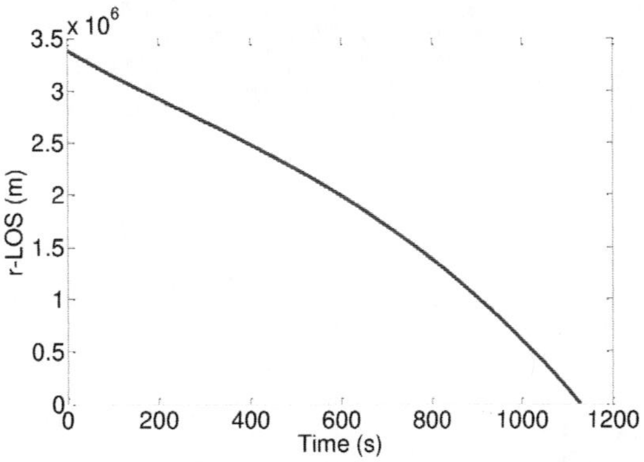

Figure 6. The radial distance between radar and the target.

4.2. Translation compensation and micro-Doppler extraction

Suggest that the carrier frequency is 10GHz, the pulse recurrence frequency is 1000Hz. The white Gaussian noise is 10dB. The precession angular velocity is $\omega_v = 2\pi$, the precession angle is $\theta_v = 15^\circ$, the included angle between LOS and the precession axis is $\gamma = 130^\circ$, the initial phase of the target is $\phi_0 = \frac{\pi}{2}$. The coordinate of the three scatters are $P_1 = (0, 0, 1.6m)$ $P_2 = (0, 0.2m, -0.4m)$ $P_3 = (0.2m, 0, -0.4m)$. The scattering coefficient of the three scatters are $\sigma_1 = \sigma_2 = \sigma_3 = 1$. According to equation (17) and (18), we do stepwise compensations on residual translation, until $\max\left(\left|\frac{\Delta_{i+1}}{t_t(i)}\right|\right) \le 10^{-5}$. The simulation experiment shows that, the required precision is obtained after three times compensations.

After removing the residual translation by RTC method, the micro-Doppler time-frequency spectrum can be obtained by using STFT on the final compensated radar echo. The spectrums before using RTC method are shown in Figure 7(a). The spectrums after using RTC method are shown in Figure 7(b). Compared to Figure 7(a) we can find that the micro-Doppler spectrums have been recovered from the effects of the high velocity translation.

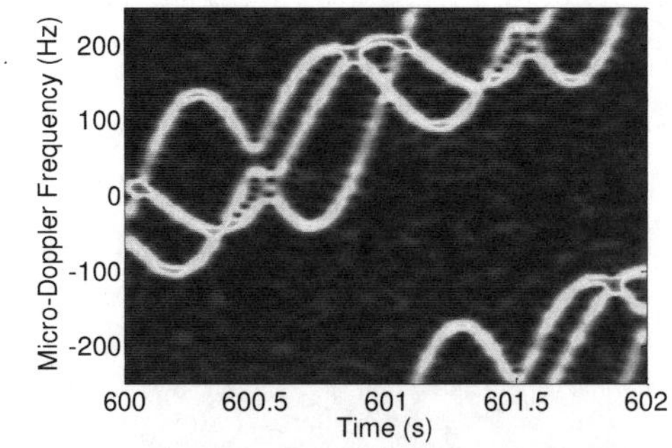

(a)

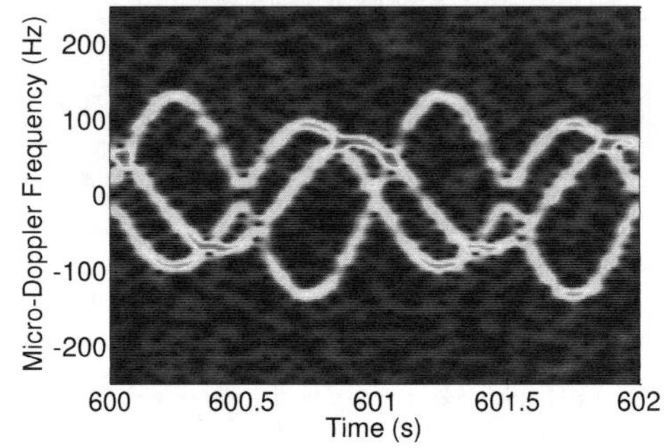

(b)

Figure 7. Micro-Doppler spectrums: (a) with the high speed velocity; (b) after using RTC method.

4.3. Tomography results

Figure 8(a) shows the tomography result of the radar echo without high speed translation. The tomography result of the radar echo with high speed translation is shown in Figure 8(b), translation causes the scattering centers defocusing. The tomography after using RTC method is shown in Figure 8(c). There are three scattering centers in Figure 8(c), which are corresponding to P_1 , P_2 and P_3 . Compare with the result in Figure 8(a) and (b), we find that this method can eliminate the translation effects. The coordinate of the scattering center is three-dimensional, so that the scattering centers in the planar tomography are the projection of the three scattering centers in radar LOS. The tomography result confirms the validity of the proposed algorithm.

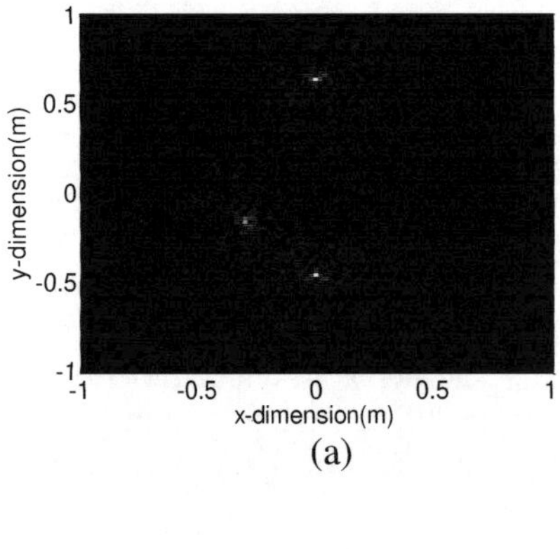

(a)

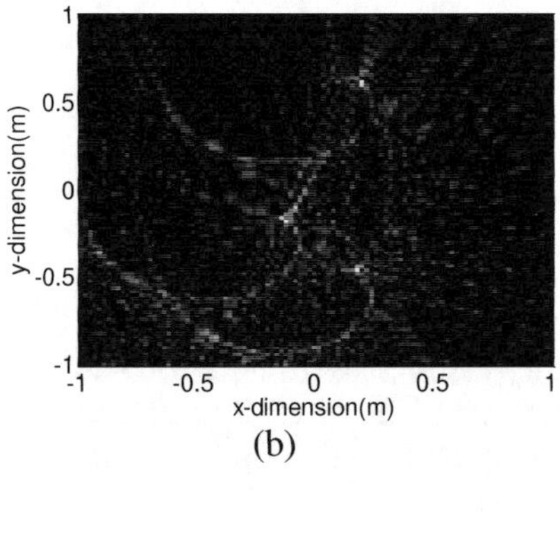

(b)

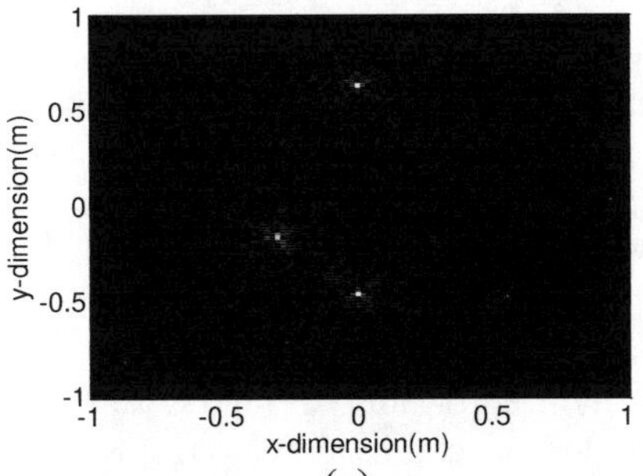

(c)

Figure 8. The tomography results of the radar echo with a SNR of 10dB: (a)without the high speed velocity; (b) with the high speed velocity; (c) after using RTC method.

Take P_1 as an example, the micro-Doppler amplitude without translation is 132.8797Hz. When the estimated micro-Doppler amplitude is between $(-133Hz, 133Hz)$, the micro-Doppler can be estimated exactly. Figure 9 shows the accuracy changes with SNR. When SNR=20 dB, the accuracy of the estimated micro-Doppler frequency reaches up to 90.40%.

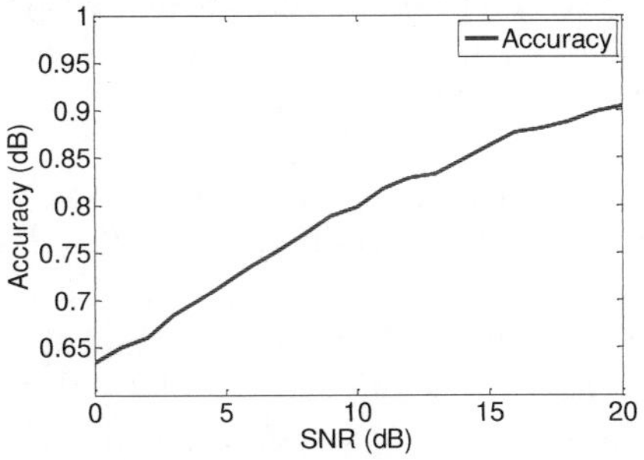

Figure 9. The accuracy of estimated micro-Doppler changes with SNR.

4.4. Comparison

The spectral peak based translation parameters estimation method designs a quadratic equation to model the translation, and then uses the spectral information of the conjugated multiplied signal to figure out the velocity and the acceleration [3]. The translation can't be simply modeled as a quadratic equation, so that the estimated translation parameters have large deviation. As is shown in Figure. 10(a), when SNR=4dB, the scattering centers can be extracted clearly by RTC method. However, the scattering center in Figure. 10(b) is defocusing by method in reference [3]. So we can conclude that, RTC method in this paper performs better than method in reference [3].

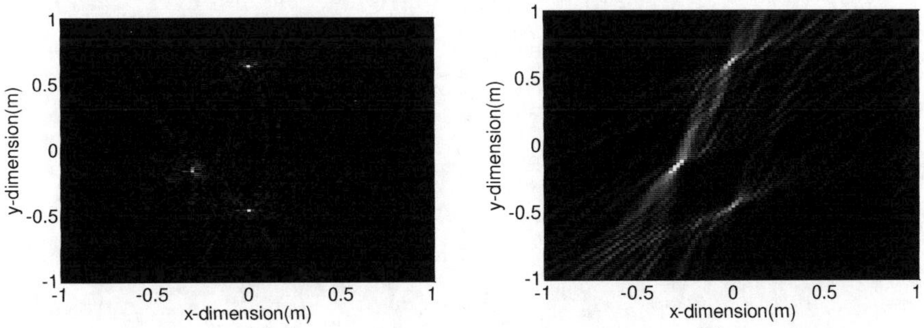

Figure 10. The tomography of the radar echo with a SNR of 4dB: (a) RTC method; (b) Method in reference [3].

5. Conclusion

Aiming at eliminating the influence caused by the high velocity translation, this paper proposes a Residual Translation Compensations (RTC) method based on trajectory information. We establish the two-body orbit model of the radar target, and then build radar echo signal model of the micro-motion target. After that we use the trajectory information to conduct stepwise compensations on residual translation until the required precision is achieved. The uncontaminated micro-Doppler spectrum can be obtained by using Short-time Fourier Transform on the compensated radar echo. Finally, the tomography results confirm the validity of the proposed algorithm.

Simulation results demonstrate that the micro-Doppler signature can be estimated accurately, and the scattering centers can be extracted clearly in tomography results. When SNR of the radar signal is 20dB, the accuracy of the estimated micro-Doppler frequency reaches up to 90.40%. Moreover, the tomography result shows that the scattering centers can be exact clearly with a low SNR of 4dB.

6. Acknowledgment

This work was supported by the National Natural Science Foundation of China under Grant 61701510.

7. References

[1] H. W. Gao, L. G. Xie and S. L. Wen, "Research on the influence of velocity on micro-Doppler and its compensation," Aerospace Electronic Warfare, vol. 24, 2008, pp.46-50.

[2] Y. C. Yang, N. N. Tong and C. Q. Feng, "Translation compensation and micro-Doppler extraction by using the information on the strongest scatter," Journal of Xidian University, vol. 39, 2012, pp.147-153.

[3] S. S. He, H. N. Zhao and Y. S. Zhang, "Translation motion high accuracy compensation for procession ballistic target in midcourse," Journal of Electronics and Information Technology, vol. 38, 2016, pp.1-6.

[4] X. Q. Luo, C. J. Shan, G. Z. Lv and J. E. Yu, "An efficient compensation method for ISAR imaging based on short time adaptive Guassian chirplet decomposition," IEEE Workshop on Advanced Research and Technology in Industry Applications, 2014, pp.891-894.

[5] H. W. Gao, L. G. Xie and S. L. Wen, "Research on the influence of acceleration on micro-Doppler and its compensation," Journal of Astronautics, vol. 30, 2009, pp.705-711.

[6] X. N. Xi, W. Wang, Y. D. Gao, Fundamentals of Near-earth Spacecraft Orbit, Publishing House of National University of Defense Technology, Changsha, 2003, pp.88-112.

[7] X. Zhou, "Study on generation of trajectory parameters for ballistic missiles", Xian Univercity of Electronic and Science, 2014, pp.24-26.

[8] M. Cao, Y. X. Liu, X. L, "Study on orbit determined methods for ballistic target", Electronics Optics and Control, vol.11, 2004, pp.35-37.

[9] V. C. Chen, F. Y. Li, S. S. Ho, "Micro-Doppler effect in radar phenomenon, model and simulation study", IEEE Transaction on Aerospace and Electronic Systems, vol. 42, 2009, pp.2-21.

Institute of Physics
Dirac House, Temple Back
Bristol BS1 6BE UK

ISSN: 1757-8981
ISBN 978-1-5108-6282-1

CURRAN ASSOCIATES INC.
proceedings.com